# Black Widow
# White Horse

Carmine B. Littleworth

# Dedication:

Dedicated to Issac

# Table of Contents

# CHAPTER ONE

# The Dogs

The dogs didn't bother the older woman that much until the day the young black one came home from work mid-way through the Covid summer of 2020 to announce that the Humane Society had called. They wanted her to bring her dogs in to the adoption shelter the next day. Mari Smith, the young one, was half black and half Hispanic. Her all black husband, Cedric, had been murdered by Boston Police less than a year before. He'd been having a mild PTSD attack after serving two tours of duty in Afghanistan. A white neighbor called it into the cops. It was referred to as a domestic disturbance.

The older than middle aged white-privilege woman, Jean, had pseudo- adopted the young black couple because they were, reportedly, being mistreated by one of Jean's man pals from up north. The young folks were stuck out in the middle of nowhere with no way of getting into town. It was the man pal's mother's ranch two-and-a-half hours north of town. That's what Jean was told when the young people showed up with their trailer-load of stuff, and their large un-neutered Chihuahua. They wanted to stay with her for a while until they could figure out their next employment opportunity. When she had offered to help them she didn't realize it would mean sharing her house that she had been hoping to remodel over the winter.

Jean Stinson Lloyd knew she was white privilege when she first heard the term long before it became a popular discussion back in the late teens. She was a true white baby boomer through and through. A California born, tree hugging, college educated, home owning, blonde baby boomer. Back in the day both of her white-collar-working-class parents would leave her in charge of the house throughout the week to get herself off to school, lock the door, wear a key around her neck, and say good-bye to the family dog. In the afternoon the reverse was her childhood routine from the age of seven.

When Mari announced she was taking her dogs (by this time there were now two Chihuahuas) to the shelter for adoption, Jean plugged her ears and started to cry, "I can't do this." She was upset about Mari getting rid of her dogs because in her family growing up, dog was God spelled backwards, and you didn't get rid of your dogs. Kids maybe, but not dogs!

Jean, eavesdropping on Mari loudly laughing to her new Istanbul, Turkey, Iranian refugee, Christian, boyfriend over the telephone, heard her say to him, "I can't do this anymore!" She had mimicked the old white woman's exact words. Mari was probably frustrated with Jean who no longer was being supportive of the young black widow's agenda. What Mari didn't realize was the same thing Jean didn't realize either at first. Jean's only friend from first grade on had been the family dog. Perhaps if her parents would have treated their daughter as well as they had treated their black miniature poodle she might not have turned out with the tragically low self-esteem that had conflicted her sorry white ass for the past six and a half decades.

Mari, on the other hand, grew up poor but both of her parents had been totally devoted to their children. She grew up in a tight family with brothers and sisters all living together in a one-bedroom apartment. Her dad was a military black man and her mom was a Mexican woman who had never worked outside the home a day in her life. There were times when Mari was growing up that her family had experienced homelessness. She remembered giving her subsidized school lunch away to friends who would then take the food home to their families that were far worse off than she. When Mari related her history to Jean the two women were discussing how much food was being wasted in the world. Jean could easily see how having too much caused waste, but she didn't waste, she was a keen environmentalist.

It was true, Mari's family didn't have much, but they did have each other. There was no way she was going to understand how ridiculous Jean was acting about finding new homes for the two dogs. The two women simply came from vastly different backgrounds. Jean got love and nurturing from a dog and Mari from her parents.

One had money, shelter, food, and education but no parents on duty. The other lacked the white-privilege basics, but was idolized and well loved by her family. Jean ate shit in many of her relationships because she was desperate to be loved. Mari was the princess that expected to be treated like one. Jean didn't know how to have boundaries because she wasn't allowed to say no to her psychologically challenged mother who had found her father hanging from the end of a rope at age ten. Mari, on the other hand, knew exactly what she wanted and thought nothing of going for it hook, line and sinker. She wasn't drop dead-gorgeous but attractive with a great smile. She knew how to put herself together, and had a knack of making everybody else's shit stink but hers. She liked to play innocent, but it was clearly a well-rehearsed act to get what she wanted in life. She knew how to play the game and play it well.

Jean was attractive when she was younger, one of those natural, outdoorsy, athletic types. As an older woman she could still turn the occasional head of a younger man, and the old guys were regularly enchanted. She still had long shapely legs and because of her outdoor background, hard work at the end of a shovel, she maintained a very shapely bust-line. She fit in with the men really well. She could talk shop or just listen with sincere interest when the guys were enjoying their favorite subjects like tools, war and mechanics. She just liked listening to the men being men. For Jean, there was psychological safety in knowing that a man thought differently than a woman because she found women to be, more or less, the unfriendly and dangerous sex.

Of course, there was nothing wrong with Mari knowing what she wanted and going for it. Nothing whatsoever unless you stopped cooperating with the "Turkish Agenda," which was what Jean began calling Mari's new life plan for self-advancement. That's what happened between the two women when Mari announced that she'd fallen in love with a man seven years younger, who was living in Turkey, who had an amazing job, and who wanted to get married and have kids, and become world philanthropists because his family had money. Jean could not deny Abdul was a good looking guy. Mari had shown his photo back when the women were still talking. He was tall, dark and handsome. Mari loved tall men. Cedric, the dead husband, had been short and didn't want to have kids. He was gone now so it didn't really matter. Mari was making a new life for herself, and her two dogs simply had to go.

# Desperate To Be Loved

The Covid Summer of 2020 played out as a global horror story. The virus quarantine was a major inconvenience. Simultaneously, fires were engulfing the West Coast for another smoke filled summer. Jean, who had always lived in the West, became accustomed to having no fresh air to breathe during fire season. She didn't totally hate fire season because it forced her to stay inside to work on her indoor projects. An artist, writer, and seamstress who would normally get up early to get her indoor work done, but when fire season came it made it that much easier. Everyone was stuck inside both day and night.

The solar landscape lights that lit the path between the two houses were failing to shine at night. The smoke during the day shaded the sun. That was the best part of the smokey summers. The temperatures were lowered by as much as twenty degrees, thanks to the big ball of fire being obscured by smoke. Jean dwelled incessantly on God's amazing creation, and wondered if fires would help to cool the planet over the next million years or so. She was a plant and planet person, and like so many, very aware that forest fires were killing the lungs of Earth. Over time, forests would regenerate, especially if humans were to became extinct. Science and God were synonymous for Jean, and with fire being the great cleanser, she stood in awe of life on earth 24/7. Mari's spirituality was far more fundamental.

Mari had been stranded on the East Coast after Cedric died. She reached out to Jean for help. Jean could never say no. She rode in on her partially lame, grey-in-the-muzzle white horse to save the day. Mari and Jean became quarantined together at the newly remodeled home needing only a few more finishing touches. John, Jean's ex-husband, lived next-door in the north house. He easily could have worked to finish up the south house had Mari and her two barking dogs not turned up.

John was a white East coast transplant who didn't do small barking dogs. Jean hoped to finish the remodel, get it on the market, and then spend her time up at the family cabin in the mountains. She had inherited it after her parents died. Now she had three houses. She divorced John before her mother passed away. John was trading work for rent. It kept him active, healthy, and in-shape without the stress of property ownership. That stress was left to Jean.

As the older woman walked back along the dark path that first night after the dogs were gone she breathed a shallow sigh of relief. Her plan was coming together. When Mari first arrived from the East Coast the two male dogs went around marking their territory in her newly remodeled home. Jean had just spent $50,000.00 on the house re-model for re-sale purposes, and enjoyed a whole twenty-six hours of semi-completed newness before she flew to Boston to drive Mari back to the West coast. When Jean saw the dogs tagging and flagging with their liquid markers she told Mari, from day one, they had to go. That was seven months earlier when the pandemic was just getting a toe hold. There was no way she was going to allow the destruction of her investment dollars.

She was feeling squeezed. Her retirement plans had been rearranged by God. When she flew back East to drive Mari out West she was figuring that the young woman would: get a job, get an apartment, and be off on her own in a couple months or so. With Covid, things came to a screeching halt. There were no jobs to be had. Everyone at "the compound, "which is what they called the two houses on three acres in horse country, went into chill mode. Jean and John found themselves enjoying the isolation and down time that the pandemic provided. She was finally able to work on things she had been ignoring for years. Now there was plenty of time to get the vegetable garden ready for food production without the unnecessary distractions that trips into town inflicted. She didn't realize that while she was enjoying her down time, Mari was busy trying to score a new man on the internet.

It was two days before the pandemic went full bore that early March when Jean went back East to fetch Mari. In less than a week they were back home in Mari's relatively new, bright blue Jeep Cherokee pulling a U-Haul trailer. The timing was profound. If Jean were to have stalled her departure a couple more days, things would have had a different outcome. Mari would have been quarantined on the East Coast instead and this story would not be told.

Jean did all the driving because Mari couldn't pull a trailer or drive in traffic. They made it across the country in four days with a hotel stay for two of the nights. Jean was fighting not only the pandemic, but also a very pricey expedition since she'd picked up the tab for the U-Haul and the hotels to boot. At least Mari bought and pumped the gas. Jean actually felt guilty cutting the deal with Mari to purchase and pump gas. Mari had never had to pump gas before. She'd always had a man. Cedric. They'd met when she was fourteen. Now he was gone, and she was stuck with this old white woman who, fortunately for her sake, believed in fixing America's civil rights problems once and for all. Four hundred years of bullshit put the United States of America looking like fools in the eyes of the world. Deemed as one of the top ten racist nations on Earth according to the United Nations, Jean felt it her duty to maintain a watchful eye on the problems that be.

As a third grade student Jean's teacher wrote on her report card that she was a natural born leader, if only she could learn how to tend to her own business. Here she was, fifty-five years later, once again snared in a situation that depleted her own resources, energy and peace-of-mind. Her old white horse wasn't completely lame yet. It was, however, getting grey in the muzzle and sick-to-death of itself. Her addiction to doing good in the world was the snare. That's what always hooked her. She wanted to make the world a better place which meant she got involved in things that were not necessary. She even told her real daughter if things got too weird she would call the family legal counsel and hand over Power of Attorney. She felt that her choices lately had been too expensive and probably inappropriate.

The funny part was that Jean knew she had a problem. She felt mentally ill from childhood. She was convinced her mother had hated her from conception because she didn't want another child. Her mother had been a career woman in the fifties after the war years, and had zero interest in being a stay-at-home mom. Even though Jean was loved by many she still felt hated. She saw on Dr. Phil many years before about a guy who loved riding in on his white horse to save the day, who ended up marrying a woman with four kids, three of whom were retarded or challenged as most prefer to say. After a year of marriage, he realized he had totally fucked up, and destroyed his life. He was addicted to riding in on his white horse to save the day. Jean knew she also had that problem. She was addicted to the rush she got when she was able to ride in and save the day in another person's world. That's when she felt loved.

The bright blue Jeep, which Mari and Cedric purchased before leaving for the East Coast a year and a half before Cedric's untimely death, had left Mari saddled with an astronomical car payment. When Cedric was alive they could barely make that payment, eat, and keep a roof over their heads. After the two women and the two dogs made it back home safely they learned that the government, due to Covid, had issued a temporary order of forbearance on all outstanding loans because a lot of people were out of work. This meant Mari's car payment could be forgotten about for a few months. It was early spring, things were warming up and greening up. The two women with John next door all settled in comfortably to the quarantine lifestyle.

Mari applied for and received a temporary respite on her loan payment. This gave the women a chance to run around and search for another lender. A better lender. A cheaper lender. First they went to a random car-dealer to find out what the Jeep would be worth as a trade-in. Then they went straight to Jeep to set up for re-financing, which is what Mari and Cedric had been told to do by Jeep when they first bought the brand new car. They were told they could refinance after one year of a good strong payment history. When Mari tried to approach Chrysler on this they refused to speak to her even though she was standing right in front of the man's desk. It never dawned on Jean, until years later, that the reason they had refused her was because her car payments had not been made on time. Jean was thinking, instead, that the system must be rigged against blacks. She believed Mari had done everything right and had been another victim of white supremacy. It never dawned on her that Mari and Cedric had been late on their car payments. If the dealer would have disclosed this information to Jean maybe she would have chosen a different path. Years later, Jean would remember back when Cedric was alive that Mari had a problem with going to the store and buying useless crap. All the red-flag warnings were there.

Carmine B. Littleworth

It was the auto dealer Carmax that revealed to the women that the car was underwater. The Jeep salesman didn't have the balls to tell the truth. Mari owed thousands of dollars more than the car was worth. If she were to trade in the Jeep for a cheaper car she would still be paying on the Jeep.

Jean was flabbergasted at the nature of the predatory lenders! In the back of her mind she kept wondering if her friend was a victim of systemic racism and treated differently because of her color. Jean went home that night to tell John. The first words out of his mouth were, "It's because she's black." Both John and Jean felt that black people were treated unfairly. They were taken advantage of by setting them up with an astronomical car payment that would eventually end up in default. They would pay $43,000.00 for a $32,000.00 automobile that was only worth $17,000.00 two years after the purchase with $36,000.00 still owing. Mari was screwed. The women went home that night dejected. Each went to their respective bedrooms to ferment in thought or maybe Jean was the one doing all the thinking, and maybe Mari just got back on the internet to trawl for her next man.

Jean and her white horse couldn't stay out of it. She had to ride in and fix everything. She got off on it. Mari, who was prone to panic attacks, was in a position to relax and get her health back in order. After the ordeal of Cedric's death Jean hoped this would happen. The fact that Mari didn't want to help out around the house was no big deal for Jean for the first few months. The young widow (but truth-be-told, a rapidly approaching middle-aged widow) liked walking her dogs instead. She lost weight, got healthy and looked fantastic. She tried hiking with Jean in the hills, but couldn't keep up with the old woman who had lived the hiking lifestyle for decades.

The neighborhood was a nice area for dog walking with calm country streets, one acre parcels and farm animals. Unfortunately for Jean, she had an easily accessible wad of cash in her savings account. She made the mistake of thinking she could help Mari get back on her feet re-building her life through re-financing her car. She felt that Mari would be very successful in the work place. After all, she was a delightful, engaging, attractive young woman who had much to offer the world. Jean offered the young widow words of encouragement because she knew that living in the white world was a challenge for people of color.

Jean ended up paying off the car loan, and then setting up a personal loan for Mari to pay back. She believed Mari to be a safe credit risk and that she would go into the world to do good things. That was one of her criteria for making decisions in life. She wanted everyone to love each other so that all people would come together to heal the planet. She didn't think it was right to destroy a planet in the universe during her life time. She was betting on Mari to succeed to make the world a better place.

Achieving a good credit score is obvious for white people growing up fiscally responsible, but for people who have not been indoctrinated over time, to work, pay, and not over extend, it has the potential to be problematic for some, black or white. Jean wanted Mari to have a chance to make it as a widow of a two-time Afghanistan war veteran. By paying off the predatory lender, Jean had cut $8,000.00 off the original loan. The new loan generated at the credit union used funds from Jean's savings account as collateral. There was no way Mari could fail. Jean declared if there was a problem making the car payment they would handle it in-house with no outsiders. This was Jean's promise based on 'her' logic. The two women went in to sign the loan papers, and the loan officer stated that Jean was being very kind. She figured that Mari had a good job, and at that time was in a stable relationship with a new, good man. She had met a local law enforcement officer on-line. Jean was envisioning that Mari would be able to pay off her car loan, build good credit and re-build her life. Both women hoped that maybe someday she would have a baby, which is what Mari said she wanted to do. Jean was positive that Mari was a good, safe credit risk.

Jean and Mari, however, had different ideas about life. If Jean had been like Mari, in the same problematic situation where someone had come to her rescue, driven across the country, paid for the U-Haul and hotels, received a free place to live with two male dogs (one un-neutered), AND put up the money to refinance her car to bring down the payments, Jean would have unequivocally jumped at the chance, every second of every day, to pay back the favor by helping out with housecleaning and yard work. This was not the case with Mari who chose instead to stay in her room, locating the man that God had in store for her, and Jean had no idea this was happening! She thought Mari was holed up in her room depressed and mourning the loss of Cedric, who was one of God's special ones. He had been forever fun, smart, kind, gentle, sincere, mellow and helpful. He served his country twice before being murdered by the police.

# CHAPTER THREE

# The Benefactor

Of course Mari's new neighborhood was white. When the two women pulled up with the U-Haul trailer in the middle of the night, Jean warned about the guy kitty corner across the street. He was a transplant from the South loaded with guns and alcohol, but he had never shown any overt indication of trouble. They were neighbors and friends at church, but Jean decided to play it Better-Safe-Than-Sorry. She drilled Mari regularly about minding her back. She probably had more fear of her white neighbors than the young black houseguest.

Jean had been a sympathizer of black America since the early seventies. She had logged hundreds of hours of black study through film and literature, but not because she was one of those white women who were hot-to-trot to taste the forbidden fruit, but rather because she'd had a bad experience with relatives in Florida as a teenager. Her own family, down South, called her a nigger-lover because she was a fan of the black baseball player, Vida Blue.

When Mari came home one day after walking the dogs, who were panting excessively from dehydration, she commented in passing that she had met a man in the park and struck up a conversation. Jean always saw the best in everyone until there was reason not to. She didn't think much of it until Mari announced she was being stalked. The stalker informed Mari he believed she was now his girlfriend. Fortunately, Mari neglected to tell the man where she lived. When he called her on her cell phone she simply blocked his number. Strangely enough he was able to reach Mari on her other phone. She told Jean that she had repeatedly let him know that they were not in a relationship. Mari laughed uproariously at the situation. She reported to her "benefactor" she felt bad that she had to tell the wacko dude, "No way, Jose." At that point Jean was starting to get a little nervous.

Mari saw herself as a wonderful Christian woman who was always ready, willing and able to serve God by being kind, but only to men. It seemed to Jean, anyway. Maybe that was just Jean's observation. When Mari tried sneaking out early one morning to meet her new-white-police-officer- boyfriend, who was picking her up out in front of the house, the hue of the color red on the warning flag intensified. That flag was now heading up the pole!

Again, Jean gave Mari the benefit of the doubt. She needed to believe in Mari because Jean's butt was also on the line. She didn't want to be forced into giving her real daughter Power of Attorney prematurely.

The two women were discussing the finer details of the new car loan. Jean was attempting to treat Mari the same as she would her own daughter by giving her a leg up in life. She would have helped her real daughter under any circumstance and she wanted the same for Mari. Jean did not want to go behind her real daughter's back by not disclosing about the re-financing of Mari's car. She just laid all of her cards out on the table and told the truth. She half expected her real daughter to raise a stink, but she didn't. She trusted her mother's judgement and went along with the deal. It's possible she may have felt proud that her mother included her in on helping out with the Black Lives Matter movement, but the black widow and the white horse were now tangling themselves into a more complicated web.

If it were to turn out that Mari was just another narcissist who would take Jean to the cleaners, then Jean might risk losing her family's wealth which had been passed to her, and which she was hoping to pass on to her own daughter. If Jean had a mental disorder, causing her to give away the family farm because of low self-esteem, or from a desperation to be loved, then she knew well enough to call her real daughter to let her know that the attorney needed to be contacted for the POA to be invoked over to her only living kin. At this point, Jean knew she could wiggle herself out of a potential financial disaster. After all, she hadn't really known the young black couple all that well. She just liked jumping in on her white horse to save the day. Only this time thing was stickier than normal. This time someone was living with her and owing her thousands of dollars.

Fortunately, or so it seemed to Jean because she always tried to keep harmony in place even if it was fake, the two women had a seemingly wonderful relationship. Jean needed to believe they were becoming good friends under a difficult situation in order to carry on under the stress of co- habitation. She was trying to promote a mother-daughter relationship in hopes that Mari would feel a sense of familial obligation. If it were a good and perfect world there would have been nothing to worry about. Both women, however, had trust issues from childhood.

The white Italian cop that Mari had found on the internet was far from evil. He turned out to be a decent guy. He had issues like a crazy ex-wife and custody of two teenage daughters, but theoretically there was nothing wrong with white police officers. Everyone who knew the white horse/black widow duo seemed convinced that the cop who had murdered Cedric surely must have been a white male, and it was Jean's white friends who were saying this. If it wasn't for Covid, Jean might have minded her own business and ignored Mari's early morning departure from the compound for that first date with the white cop.

She listened closely hearing Mari prepare for her crack-of-dawn-departure letting the dogs go out for a pee, quietly using the bathroom, and then slipping through the telltale squeaky front door. Jean laid in wait. Mari made it about twelve steps out when the benefactor, who was sitting by the sliding glass door of her art studio popped her head out to say, "Hi, Good Morning, will you be gone today?" Mari was caught. She should have just said, "Yes! I have a date." That would have signaled to Jean to back off. But the undeniably, complicated relationship between the two women made Mari feel like she had to answer to Jean since she was 'given' a free place to live.

It freaked Jean out that Mari was fraternizing with a stranger during the pandemic. She didn't try to stop Mari from going. She acted polite and cool, but when she went next door to visit John to explain the situation he went ballistic. It wasn't necessary for her to react strongly. John did it for her. He was originally from New Jersey, and was good at freaking out. That way Jean didn't have to be the heavy. She didn't have to lose her sweetness-and-light position. She didn't like for butter to melt in her mouth unless there was no other option. She enjoyed her cool-as-a-cucumber persona. It was like she tried to keep her bitch mode under wraps until there was no other option.

Mari returned a few hours later. Jean let her know things weren't working out. She informed her that she should relocate ASAP. Running the risk of Covid at the compound was not an option. Jean felt threatened that the young black widow would begin dating regularly. She feared that a parade of male suitors would soon be wafting through. Jean issued a mandate for the compound: No Visitors in 2020. Not even her daughter and son-in-law were welcome to come back for their yearly vacation home with mom in August.

Soon, Jean was flapping her jaws to her close circle of friends. She told them Mari was dating a white cop. It wasn't so much that he was white, or a cop, but the fact that Cedric had barely been gone seven months that raised an eyebrow or two. Everyone in Mari's relatively new group of white support had been very fond of Cedric. He was exceptional; he just was. Jean almost fell over backwards when she heard the news. " A white cop! Are you out of your mind?" The two women giggled at the reverse discrimination. It appeared Jean was almost more prejudice against white people than Mari. To make matters worse the cop was an Italian!

Jean was no virgin to the game of dating and love. She'd had plenty of both. She reflected back on her past two hot Italian lovers, and felt the need to tell Mari, "The problem with Italian men is that the entire relationship is based on the man's penis! Look at the Italian art, take Michelangelo's, David for example!" She grabbed her art book to prove her point. She showed Mari a photo of the statue. They both fell into uproariously loud laughter. Jean wasn't really prejudice, she just liked being theatrical if it were something she felt strongly about. She didn't want the young woman to fall further into harm's way. The nasty white world was something that Jean thought Mari needed protection from, but she had no idea what she was doing when she did it. She just wanted to shoot the white horse dead once and for all.

# CHAPTER FOUR

# The Biological Clock

Everyone knows that when a woman still wants to have a baby and her biological clock is ticking a sense of urgency occurs. Fortunately, the Italian turned out to be a very nice man even though he was a few years older, divorced with kids, and a troublesome ex-wife. His patrol partner was a white man with a black wife. Mari enjoyed socializing in the new foursome. John and Jean never got comfortable with Mari fraternizing with outsiders during the pandemic quarantine, but it didn't matter. Mari assured them that everyone was being tested for Covid weekly.

Mari worked temporary jobs until landing the perfect employment she had been hoping for at the Regional Medical Center. The pay was good, with benefits and advancement. It appeared she and the cop were a nice respectable couple. He worked for the highway patrol with five years left until retirement and a willingness to start another family. It looked like everyone was settling in to a blissful life where Mari would be able to pay off her car loan to Jean, or so she was hoping.

Carmine B. Littleworth

With expectations being the easiest way to set yourself up for a disappointment, Mari announced she'd found another man online, and was now going off to Istanbul to meet him. Jean realized that being paid back for the loan on the Jeep might come with a couple of warts and freckles. She told her friends if she could not afford to lose the money then she should not have taken the risk in the first place. Unwilling to set herself up for a self-induced screw job she refused to be obsessed about losing the money. She had to take the position that she may have lost a bunch of bucks and so-be-it. She had worked hard to help the not-so-young black woman get re-established legally with vaccinations for her dogs, vehicle registration, car insurance, driver's license, and even state-sponsored health insurance with an EBT card. She helped her file her back taxes and deal with the IRS. She really did try hard to make her a welcome part of the family the best she could in her version of a white society's value system.

Mari stated she felt white people liked using her for her blackness. At first Jean thought that it must be a bad thing. But after Mari started work at the hospital as the front line receptionist with an unruly public, Jean soon realized that the white establishment was using Mari's delightful blackness as a measure to heal Black-White relations in America. The year 2020 had been a super doozy; between politics, pandemics, and people getting shot in the streets, and in their houses. Jean maintained, "Life on earth does not have to be so fucked. It easily could be really sweet if only the white people could learn how to chill." The question still remained, was Jean throwing stones in her own glass house? It was a very tough year for the human race.

The black one had lost her husband to police violence, and the white one got in over her head with the whole save-the-day-on-the-white-horse routine, but… she was determined to get herself out of the predicament smelling like a rose, and not fuck up again!

Mari announced she was applying for her passport to go meet the man that God had for her, and then things changed drastically. Mari's white cadre team all fully agreed that the Turkish Agenda, Abdul, must either be a sex trafficker or an organ harvester. Either way the consensus was that if it were too good to be true, then it probably was too good to be true. But Mari would hear none of it because this was the man that God had for her!

Supposedly, this Abdul character was a highly successful international businessman at age twenty-nine. His family fled Iran in 1979 as refugees, to re-settle in Germany. Mari was thrilled that he came from a good Christian family. Mari and Abdul liked reading the Bible together on FaceTime in the outdoor chairs that Jean decided to put away for the winter earlier than normal that year.

The new couple also enjoyed praying together. This was the foundation of their new relationship. They met each other in a Christian Facebook chatroom. Mari was under the impression that she was going to Turkey to meet her new man, and become a world traveling-philanthropist, while helping new her husband give money away to good causes. What could possibly be a more delightful twenty-first century fairy tale than that? And dreams really do come true sometimes! The stable cop relationship that Mari enjoyed for a few months had not met her needs for some reason, and Jean was bummed to see the white cop go.

If you were a comfortable, middle aged, white privilege woman who had worked hard all her life to see this fairy tale unraveling right before your eyes, skepticism might cross your mind. But if you were a not-so-young, clock ticking, desirous of a decent life, Black-Indigenous-Person-Of-Color, BIPOC woman, you could easily be tempted to believe that miracles really do come true. Jean prayed that Mari was right, and that she was wrong. Mari reported that Abdul wanted to buy her a first class plane ticket to Istanbul so they could begin their new life together. Jean's old and pragmatic brain chewed on that one for a few days. Out on the driveway one morning, before the smoke-filled part of the summer started up, Jean downloaded onto Mari's ears, "Make sure you get a round trip ticket. Do not leave this country without a round trip ticket!" Then she proceeded to deliver her logic. "A one way first class ticket to Istanbul is $11,000.00. A round trip ticket is $8,500.00. Why is a one-way ticket more expensive than round trip?" That was red flag number two. Secondly she researched the coach fares. "A round trip ticket is $900.00, and one way is $800.00. For an extra hundred bucks it's cheap insurance to pay, up-front, for a way home in case things don't work out exactly as you planned!" Jean was always stressing to Mari, "Never do anything without having a plan B always available for Back up."

After Jean delivered her request for Abdul to purchase a round trip ticket the women went back to their daily routines. The thing that Jean feared the most was Mari getting stuck in Turkey and that the first person she'd be calling was the old white money bags back in America to come to the rescue. Jean had already paid well over $30,000.00 towards the Mari project and that was enough. She had cost her more than her own daughter and her cousin down in Florida combined. It was time to heal her mental white-horse-illness once and for all.

Mari texted Jean in response to the suggestion of a round trip ticket. "< Oh, Abdul and I have other plans. I won't need a round trip ticket. We plan on taking a little vacation.>" At that point Jean just shut down. "Okay, cool", she thought, over the loud speaker that was, finally, blaring in her head.

Jean was not mad in the slightest, although Mari thought she was. Jean was resolved to get out of the mess she had created for herself. It was the perfect segue into giving her the freedom-out, for which she had been patiently waiting. She simply disappeared from Mari's life, even though they were living with a common wall between the two bedrooms.

Jean's room had a sliding glass door exit. She could come and go over to the house next door without detection, or at least very little. She was practicing avoidance. After a day or so it had become obvious that Jean had decided to protect herself. It was more than the money. It was healing a lifelong relationship of always getting screwed by setting herself up for the screwing. She was letting the white horse go. She was opening up the gate to the big grassy pasture so the white horse could retire without her. They walked to the gate together. She took off the bridle. She opened up the gate. She let the horse go. Bye-bye white horse, and the white horse with the sore back was glad to be done with Jean as well.

The two women continued texting with each other the basic perfunctory bits and pieces. Mari did try to ask if they could talk. Jean felt it was unnecessary. This went on for two and a half weeks. Finally, on Mari's day off from the hospital she called out to Jean who was ready to leave to go into town, "Jeannie?" "Yes?"

The two women stood in the hall talking for two and a half hours sorting out the crap. For Jean, things had not changed. Mari was on her own. Jean was thrilled. For Mari, she was bummed that she had lost her resource supply and access to her Mama Jeannie which was the name Cedric and Mari had given to Jean back when they were all living together before the murder. Back before the couple had relocated themselves to the East Coast in their brand new Jeep. Jean had loved that name for years. She felt like she was helping to heal America's race issues. But more recently she was feeling like she had become the sugar mama instead. Jean had always been plagued with the mental illness of not feeling loved. Even if Mari really did love her she couldn't feel it anyway. She was desperate to be loved and went to great lengths to get people to love her only to get herself into situations that compromised her lifestyle.

Mari made the comment, early on, about white people using her for her blackness and Jean pondered on this for a long time. She was obsessed with healing the racial divide. Jean thought to herself, What's wrong with the hospital putting a pretty, young woman of color with a great smile as a first responder to the public, at the front door? If white America needs to get used to the idea that Black Lives really do Matter, then it must be a good thing. Was Mari really justified in feeling used? On the other hand, what's wrong with a black woman using a white woman for her whiteness? Jean was feeling, Donald Trump should be giving her a tax credit for helping out with the reparation payments that the black people are asking to receive. After all, the government paid reparations to the Japanese a few years back after taking their property, and then sticking them in an internment camp for the duration of WW2!

As the two women talked in the hall, it came out that Mari felt hatred coming from Jean. After all, she did just disappear on her. She simply vanished from Mari's life even though they were sharing the same house. This was the two and a half weeks Jean needed to get some space, and for Mari to fully transition to her own two feet. But Mari felt hatred. When she told Mama Jeannie how she felt Jean's response was, "Hate? You feel hate? After all I've done for you, you feel hatred?" She wasn't angry, just shocked. Jean explained that she needed to protect herself. She needed to stand up for herself. She had never been allowed to have needs of her own growing up. She was only allowed to do exactly as she was told. Her mother would tell her when she was little, "Yours is not to reason why, but instead to do or die." As a child she would think about this on a very deep psychological level. Simply do as you are told or you'll be dead. But in the situation with Mari she felt like she would be dead if she continued doing as she was doing. She had no idea where it would stop, and she had to stop in order to survive. Jean could certainly see how Mari felt, but it didn't matter.

It was forest-fire-smoke season toward the end of summer. Jean felt the need to get ready for winter early that year. She took Mari's favorite outdoor sitting places and removed or changed them in such a way that they were either gone or relocated, but clearly unavailable. She put the beautifully framed charcoal drawing of Mari and Cedric, that had been lovingly displayed in the living room as a tribute to the fallen soldier, back into Mari's room on top of her bed while she cleaned the floors of the entire house. She told her to clean her bathroom. She did all of these significant gestures to let her know she was to be gone by a certain date because Mari said, back when they were still talking, as soon as her passport arrived she would be leaving. Jean didn't enjoy being passive-aggressive, but she had clearly maxed out on Mari.

Poor old Jean had bent over backwards to accommodate the young black widow only to have her jump ship with a man she had never officially met- face-to face. Mari's plan of taking an international vacation during a pandemic, only to leave Jean holding the bag, was the straw that broke the camel's back. There were so many unanswered questions to consider. Because Jean could not bring herself to continue further involvement, she just shut down. Not a word was heard. This allowed Mari a chance to plan for herself. This allowed Jean a chance to take a giant step backwards and get some space to analyze her own behavior. She kept saying over and over to herself. I will make it out of this one alive. I will make it out of this one alive.

At the end of the two and a half weeks they were back on speaking terms. It was good they both had some breathing room. They were able to get stronger in their own individual endeavors. Jean told Mari she needed to be able to move on with her own life because she felt like her own life had been put on hold. Hijacked was the description she used. Covid was nobody's fault, but it definitely threw a monkey wrench into the cog. The time had come to move forward for both parties. The separation gave the women an opportunity for appropriate boundaries. Mari had her own life to live and so did Jean.

Things went back to normal. They were talking. Jean stopped calling Abdul 'The Turkish Agenda' under her breath. She never did say it out loud. The new young "international" couple continued carrying on with their plans. The separation between the two women had enabled Mari and Abdul to make their own plans without Mama Jeannie's involvement. This made Jean very happy. Soon all of Mari's stuff piled in the garage would need to go to Salvation Army. The Army would soon be sending a truck. Mari would wait for her passport to arrive. She asked Jean if she could stay on for a while longer until everything fell into place. That was not a problem.

# CHAPTER FIVE

# Mother Nature

And so as it was that the two women began discussing babies with great joy. Mari would need to get pregnant by the end of the year just before her 37th birthday in January. It appeared God was coming through for Mari and bringing her a good man. The new young couple had their strong faith in the Lord which was the same reason Mari and Jean were able to carry on their close friendship. Mari's parents, Mari's new boyfriend/fiancé and Jean all had one thing in common: Faith. It didn't mean they stopped acting like normal people, it just meant that when all else failed you could always trust in the will of God. Jean was not able to fault the young couple for their similar devotion to her own basic belief system which maintained faith in a higher power.

Jean's spirituality was quite a bit different than Mari's. Mari was basically a fundamentalist Christian, and Jean leaned more towards eastern philosophy. They never argued about faith. They never disagreed. Even though Jean was a student of the Bhagavad Gita/Hindu faith tradition she was devoted to Jesus because she felt that Jesus had come to her one day while out hiking in the hills, tapped her on the shoulder, and asked her to go to bat for the

Christians. She was a student of Buddha as well. She could not possibly deny Jesus' request and said back to the Lord, "Sure, Jesus, no problem. Bring it on bro!"

Mari loved hearing that story as the two women made their way across the country pulling the U-Haul with two yapping chihuahuas during that first week of the pandemic. It helped set their spiritual bond. The other bond that kept them connected was Cedric. Even though he was gone, Jean would not let go. It would have been the easy way out to tell the black widow, "Oh good, Mari. I am happy for you and Abdul. See you later. Have a good life. Adios!" But instead she wouldn't let go of Cedric. It was wrong what happened to her dear friend who was like a son. The police murdering of innocent people in America had to stop! Jean knew she was going out to the pasture to bring her old white horse back in from retirement for one last ride.

She made it a point to conceal her affiliation with the Hindu faith tradition. She wasn't exactly hiding it from Mari, she was just waiting for the right entry point. Jean didn't want to make her houseguest feel like she was living with the Anti-Christ. She was being very careful not to mention The Guru. Jean had a funny way of cultivating people. The Guru told her, many years back, she didn't need to do the work in the world she just had to turn up. She had no idea what that meant at the time, but as the years rolled on she gained greater clarity.

There were actually two Gurus. One was the real deal from India, and the other was her neighbor up the street. The one from India wouldn't allow anyone to donate more than one dollar per month because he would rather have a million followers donate one dollar than one follower donating a million. The American Guru up the street was attracting a few followers with big bucks. Jean appreciated the information from both, but found the one from India to be far more humble.

She knew she had some kind of role to play in the world. She had no idea what it was other than that she needed to heal the entire planet before she died. Cedric and Mari were in her life for a reason. If racism were to end, then everyone would be able to work together to heal the planet. Asking the BIPOCs to clean up the white man's mess was going to be an interesting dialog, but she knew she'd be ready for it when the time came. As a professional, Jean was convinced it would only take forty years to heal all of the rivers and streams with healthy vegetation. She also knew that it required the cooperation and commitment from all of the people on earth to get this job done. The American Guru felt that he was at the top of the pyramid. Jean felt that an apex or top dog was inappropriate at the current stage of mankind's evolution, and preferred to do God's work from an undisclosed vantage point. She preferred not to reveal herself as Mother Nature. That was the nickname John had given her. He was kidding but she took it very seriously. Many hands make small work. Everyday Jean got herself out of bed and went to work healing the planet in some small way.

The neighbor Guru thought it was his job to elevate human consciousness. Jean felt it was her job to work toward Earth's restoration. When humanity puts all of its differences aside then the restoration can begin. But before restoration comes reconciliation. That was the challenge. The evolution of life on Earth was depending on "Behind the Scene Jean". That was how she described herself to the people she trusted.

The neighbor up the street was convinced he was God, but Jean was just another one of God's little helpers. One wanted to take all the credit, and the other wanted all of life working together. Jean realized through her journey with the Indian Guru that she, George W. Bush, and a one celled amoeba all had the exact same value. How could that be? When colonies on Earth start collapsing as they are currently, it doesn't matter if you are bacteria on a dog turd or

Jesus Christ himself. This was one of her strong opinions! It just wasn't right to destroy a planet in the universe. That was her standard rant. Helping Mari bring Cedric's killer to justice was something she did on the sideline while waiting for the opening from God to begin planet restoration globally. Both Mari and Jean were waiting on God.

The senseless murder of Cedric had occurred in late September of 2019. Every time his monthly death anniversary came back around Jean encouraged a remembrance of some sort. The first month was a big deal. The two women sat down to a meditation connecting to the spirit world. They remembered him in silence then both offered a prayer out loud. They both had a good cry. By the second month, which was the six-month anniversary of Cedric's passing, Jean had developed the ritual of setting up a little alter under the charcoal drawing of the young husband and wife in the far corner of the living room. She was obsessed with her adopted black son's murderer. She wanted to know the murderer's name. She wanted to know if the murderer was Irish, same as her. She was not willing to let it go. She couldn't. She called the District Attorney's office of Suffolk County, Boston, Massachusetts. She tried to find out what was happening with the investigation. She wanted to know if Cedric would die in vain and be forgotten like so many others. Jean had volunteered to pay for his cremation because she didn't want Mari to suffer further, and she refused to have Cedric look like he died with no means. She had been through a number of family deaths already. She knew the routine. Morticians make a killing when a person is caught in this excruciating moment of time. Most cannot bear not to have what little is left. "It's a racket," some would say.

In retrospect, it might have been better if Jean had just minded her own business and stayed out of it, but that would have meant a $1,600.00 credit card bill for Mari that she would still be paying on while planning her marriage and pregnancy with Abdul. Jean, had a rather warped sense of humor. After picking up and storing so many different containers of ashes from her own family, she declared she wanted her blood daughter to stiff the mortician when it came time for her death. Jean wanted to end up in the landfill. She always said that was her final wish.

Jean Stinson, the ardent environmentalist, believed in healing the planet. Wanting her ashes to go to the landfill was the way she saw her spirit healing up all the toxic dump sites on earth. Whatever happens to unclaimed and unwanted ashes? She queried. Jean wanted to stiff the undertaker, and got a perverse delight at the thought. They had to do something with unclaimed ashes. Why pay the mortician, in the first place, if you want to end up at the dump? Her unclaimed dead body was to serve as a rebate to offset all of the ashes she had paid for over the years, including Cedric's.

She was curious enough to call the local undertaker to ask questions. She found out people with no means get their bill settled by the government and dumped down at the river after 2 years if no one showed up to claim the ashes. Then she asked the mortician, "What if the loved one was so despicable that the entire family refused to claim the cremains?" "Same thing," replied the lady mortician, "off to the river." Strangely enough the landfill, that Jean dreamed of for her afterlife, occupied the entire hillside above the river's north bank.

She was glad Mari had Cedric's ashes in the house they shared, Jean's house. When Jean had gone in to clean the floors she saw that the young black widow had set up an alter in her room with cherished photos from their past. They had been a couple for twenty-three years. The young woman was going into the new marriage with, at least, her deceased husband's ashes free and clear of debt. Jean smiled and remember when Cedric was still alive.

When the one-year death anniversary arrived Mari threw herself into her work at the hospital and volunteered for extra shifts. She didn't want to be around Mama Jeannie any more. The poor young woman had black circles under her eyes from eating too much sugar, and not sleeping. She had lost a good thirty pounds. Abdul was working on a project in Turkey and unavailable even for texting, and Jean put another call into the Suffolk County District Attorney's office.

# Murder

The old white lady had been expecting a flaming cross to show up in her front yard for over thirty years. Her lifelong passion for righting the wrongs of the American slave history came to a crescendo during that notorious summer of 2020 when a high number of police shootings of brothers and sisters made center stage around the globe. America was being called on its shit at home and abroad. The USA was designated one of the top ten racist nations on Earth. White Supremists were crawling out of the woodwork and feeling emboldened because of Donald Trump wanting to Make America Great Again. They felt he was their leader. You go proud boys! Jean snickered under her breath. You'll get yours. Just you wait and see.

For decades she tried to explain her politics to John, only to end up in a screaming match. He followed MSNBC, and she followed the United Nations 17 Sustainable Development Goals. He thought Donald Trump was an evil person, and she thought he was a necessary evil because Americans had been such bad players on the world scene, and DT revealed that embodiment. Her standard statement was, "The purpose of Donald Trump is for America to see itself."

The only spirituality the USA was good at was the Church-Of-The-All- Mighty-Dollar. Jean was spirit minded instead. It was one of the things she and Mari had in common. Though they worshiped Jesus differently, the core faith they shared was rock solid in agreement. All people of faith shared one thing in common: Believing in the will of God. Jean took a nice deep breath while contemplating her favorite lines in the Lord's Prayer, thy will be done on Earth as it is in heaven, but Give us this day our daily bread, always sent her into wondering what that might look like if people were willing to NOT take more than needed.

There by the grace of God go us! That's why Jean and Mari would always stay connected no matter what went down. They had that unspoken and spoken bond of faith in the Lord. That would never change or waiver. But why had God willed slavery? They never discussed that part.

So hip hip hooray for the KKK, scoffed Jean, who prayed for their lack of success. "My grandfather would have loved their exposure." Her mother's father, Jean's grandfather, had been murdered by the KKK back in 1933 for being a Jewish sympathizer. That was why her mother was a paranoid schizophrenic. It always seemed strange to Jean that she felt something in common with black people. The fact that it was murder was emphatically gruesome, but not overly difficult to fathom. She imagined her grandfather happy in heaven seeing the KKK exposed and Donald Trump being the flashlight into America's darkness. John could never understand her logic. Jean would quote her mother saying "Right is right, and wrong is wrong, but never the twain shall meet." Resolve took away all the pain, but murder was still wrong. She tried to convince the young black widow to pursue action of some sort so that Cedric didn't have to die in vain. Cedric's life did matter, but maybe only to Jean. It seemed, Mari just wanted to see it all in the rear view mirror so she could have a life again while she was still young enough to start a family.

She was afraid and there was a lot to be afraid of in America if you were a person of color. While they were in the garage discussing what of Mari's belongings the Salvation Army would cart off when the truck arrived, things came to a bit of a cementing point. The one year anniversary of Cedric's death had come and gone. The Suffolk County District Attorney's office was not returning Jean's phone calls. She needed to know if Cedric's murder was still under investigation or if a copy of the investigation had now been made public.

Mari was sorting through Cedric's old military gear that the Salvation Army would pick up in a couple of days. Mari confided in Jean that Cedric had been issued a very high security clearance while in the war. He had participated in covert operations. Jean wondered if the government had Cedric killed because he knew too much.

There were big differences between Breonna Taylor, the innocent young black woman who was shot in her own apartment by the police, and Cedric the United Sates Army veteran. Cedric's story was even worse. Breonna's family had just settled for millions of dollars, but Mari wouldn't budge. She just wouldn't move forward. She gave the excuse to Jean that she was afraid the government would comb through their past. She confided this to Jean while sorting through the old military paperwork and manuals.

Breonna Taylor had received justice compensation. Her family hired the attorney Ben Crumb. The case settled for twelve million dollars. The family probably got six million dollars. Jean knew Cedric would want Mari to have six million dollars. She just guesstimated that the attorney would take about half the settlement. She didn't want one penny for herself. She just wanted the United States of America to become a nation with common decency.

During the first few months of the two women living together they would sit down at the dining room table hammering out Mari's statement of the alleged murder incident. She never had given her statement to the Boston PD because she did not want to enable them to use her statement to their advantage. Mari's statement happened to be what was missing from the investigation report that the DA's office was not releasing to the public. It was becoming more and more obvious that hiring an attorney was the only way to justice. Since Ben Crumb's website touted that a person didn't have to pay until Ben won the case, this was especially enticing to Jean who had become maxed out on paying for Mari. If she were a millionaire she would have hired an attorney right from the start. Fortunately, she had the sense to know what she could and could not afford. What she really could no longer afford was the stress of being Mari's friend. Jean was maxed out, but so was Mari.

After the Salvation Army came and cleared out the garage of Mari's old stuff, it gave space for her to be able to organize what she wanted to keep. The current plan was for Abdul to come to the United States to fetch his bride-to- be. She was busy trying to condense everything into the smallest footprint possible.

Jean wasn't letting Mari off the hook. She continued, like water on the stone, to seek justice for Cedric. "Donate the money to the SPCA if nobody wants the money," suggested Mama Jeannie, "but don't let Boston PD get away with murder!" For Jean, it was blood money.

Back when Jean and John were questioning if Abdul was a sex trafficker or organ harvester they planned that she would go to meet him with one large suitcase and a small carry on. That was how the big disagreement began in the first place. Jean had suggested getting a round trip plane ticket. Their plans changed however. Abdul got a promotion at work and suddenly was able to come to the USA, meet his future wife for the first time, hook up a U-Haul trailer to the Jeep, and drive eight hours south to meet Mari's parents and siblings in the one bedroom apartment. Then they would drive back East to where Abdul would have his new USA office and residence in lower Manhattan. It was a perfect plan if only it could work. There was a shitload of balls in the air, but Jean was willing to play along and pull the thread to see if fairytales really could come true.

The differences between the two women were so profound that Jean stopped believing in the need for a man. Mari was still in the stage where her whole world was focused around needing a man. Jean resolved after three failed marriages that she was on her own in the world. She finally stopped hoping that her dearly beloved would show up. Mari and Abdul still planned on getting pregnant by the end of the year. Lots of variables were cooking on the stove and Jean knew, because she was old and had been through some shit, God would make things interesting, as God always does.

Mari and Abdul continued to spend time praying together on FaceTime. This was interesting to Jean because when the two women had left the East Coast Mari's father prayed with them as they sat in the car ready to roll. He prayed for the two women, the two dogs and a safe journey. It took about five minutes for the prayer to unfurl, but it really helped set the stage for the trip across the country. From that time forward anytime things got a little katywompus Jean would suggest either prayers from her dad or a discussion with her dad. Jean was hoping for collaboration from Mari's real family.

Jean's dad, like Mari's dad, also had been her guiding light, until he died. Deferring to a man's wisdom was ingrained in the old woman's psyche. Unfortunately, she had never been able to hook up with a man that had the same flawless judgment as her own father. It was easy for Jean to accept Mari's devotion to a man. Jean admired those old fashioned values and saw greatness in the tried-and-true system that had served mankind's evolution through time.

Then, as if to add insult to injury, right on cue, the pandemic year of 2020 marked the one-hundred-year anniversary of woman's right to vote. Jean laughed thinking there would be a big celebration of the struggle for women's rights. It was obvious that it wasn't getting any recognition for a reason. By comparison, the riots, pandemic and politics had tossed the women aside yet again. Jean wondered if it had been swept under the carpet on purpose to keep all women around the globe from uniting.

The big turmoil instead was the wearing of a mask. A month or so before the presidential election, The Freedom Fighters and the Proud Boys were all pissed off that masks were mandatory. At the beginning of summer, Jean could not possibly see how masks could become a political debate. Some people believed in masks and some people did not. How could anyone possibly not be willing to wear a mask? Jean was flabbergasted at America's abuse of freedom. The USA during 2020 was being given an opportunity to reconcile its past abuse of human rights, namely slavery, but instead chose to fight over wearing masks?

Jean fell into the Black Lives Matter movement by no self direction of her own; she didn't go looking for it, it just showed up. Because Jean and Mari had the commonality of faith they were able to Let-Go-And-Let-God when the future was too obscure to choose an obvious direction. Every time they hit a stumbling block Jean would say, "No problem. God will figure it out." But there was a problem; Jean wanted Mari to pursue charges against the police for murder. Mari was like a dead fish. She had zero locomotion. That's how it appeared to Jean. She would relate news stories to Mari about the various police infractions happening to people of color. She would try to give her all sorts of information to encourage her to get the ball rolling with "Justice for Cedric."

It took three months for Jean to get the whole story out of Mari exactly how it happened the night of the shooting. They would sit down at the table for an hour or two at a time and the story would pour out of the young black widow while Jean would type away on her old Mac Book Air. She took Mari's statement just as if it were a legal deposition. It was the one she had refused to give to the police because of her inability to feel safe. After all, she had just witnessed the cops murdering her husband. It was not easy for that story to come out. It was not easy for Jean to hear all of the gory details. Jean didn't want the story to go unrecorded, and Mari would never commit to pursuing further action. Jean simply wasn't willing to push Mari. She would nudge gently but never push, even though the case was worth, probably, six million dollars in Mari's pocket.

After the shooting, Jean was the first person Mari called. She was too afraid to call her own family or Cedric's family. Understandably, she was out of control with grief. It was a Sunday afternoon when the call came through. Jean refused to answer calls on her cell phone that were not identifiable so she just ignored it. Then a text came. It said, "<Jean, it's Mari, answer your phone, It's a matter of life and death.>" A few minutes later the call came through. Jean answered. Mari gave the news. The two women sobbed.

Immediately Jean began hunting around on the internet for a course of action. There was no course of action unless she wanted to hire an attorney. This had all happened less than one year before Ben Crump became famous for the Breonna Taylor case. Jean, at the time, could find no one who was willing to take Mari's case Pro Bono. Nobody was willing to touch it. No one had any suggestions. Jean was not in a position to negotiate any deals, and Mari simply could not speak. She had just lost her entire world and Jean couldn't find a single soul willing to help her friend, who was now all alone on the East Coast three thousand miles away.

# CHAPTER SEVEN

# The Character Witness

Mari started the long and complicated memory of Cedric's death with where they were the night of the murder. They were managers of an Air B&B within walking distance of the Arnold Arboretum. Down at the patio table below their apartment the young black couple entertained, their friends, another couple of color who also worked at the B&B. They were all saying their good-byes to each other as newly made acquaintances. Quiet chatting and laughter filled the air as Mari and Cedric told of their plans for the future up north in New Hampshire. They were to become the new managers of a different type of resort that catered up the great outdoors and scenic beauty in a homespun, elegant atmosphere.

Mari and Cedric Smith planed on getting out of the city and moving to the country. The anti-black vibe in the neighborhood of southwest Boston was unpleasant. They were ready to move on again, as they had done so many times before. Their whole married life they had been on the run. They were good at it. They had everything completely organized for moving. The military had prepared them well. Mari was the engineer and moving manager while Cedric took orders from his wife making it all happen physically with his magnificently muscled body. There must have been at least twenty of those giant plastic tubs on wheels where everything fit nicely. The one item that didn't fit into a tub was a rocking chair that Cedric had given to Mari one Christmas when he was overseas, and not able to make it home for the holiday. It was her pride and joy and she hoped to use it some day to rock-a-baby.

New Hampshire would have been an amazing location for the young couple. Mari was hoping to convince her husband they could start a family. It would have been a lovely spot to raise a child, but she ended up alone instead. The owners, at the new job site, were willing to work with Mari's difficult situation. They were a white middle-aged gay couple who had adopted two young black children. They wanted their kids to see people like them instead of just all white folks.

The mortician had orders to ship Cedric's ashes up to the new address in New Hampshire because Mari could not make the drive back down to Boston by herself to pick them up. Cedric had been the driver and now he was gone. She, also, could not drive back to Boston to give her statement to the police, or meet with legal counsel. After the killing clean up inside the apartment at the Boston B&B, her brother came to her rescue to help her make the move up North. He flew up from Texas.

Since Mari was refusing to give her statement to law enforcement Jean was worried that Cedric would end up becoming another black life that didn't matter. The young couple and Jean had become like family because the older white woman was willing to help. Jean couldn't allow the white folks to murder one of her own without a fight. She was frustrated beyond belief. It was not possible to shine this one on. There had to be a way. She told Mari, "I can't get for you… your husband back, but we can damn sure try to get you some money out of the deal." This only made Mari more devastated every time Jean spoke on the subject.

Out of frustration Jean decided to write down her own story of Cedric's murder. She wanted to capture all of the details so that Cedric would not just blow away in the wind when Mari decided to release his ashes. She wove her own tale of a lifelong passion for black struggle. It began when she was a teenager with her mother's relatives down in the deep South. She wrote the story in case she was ever called in as a character witness for Cedric. She felt she needed to channel her anger.

She had chosen the title "Two Dead Coons" because she wanted to speak to white people about their behavior. She knew she was white, of course, but she was old enough not to care. She had been silent long enough. Jean talked John into reading her story. He didn't like it. She got her cousin up in the Pacific Northwest to read it. She thought it was too inflammatory towards white people. Another friend read it and said it was a story that needed to be told. She wrote Two Dead Coons as journal of violence, prejudice and a prayer for action.

She had been expecting trouble from white supremacy most of her life. She shot her mouth off about the destructive nature of capitalism in a worldwide group chat on the Linked In social media platform back in 2009. Her grandfather had been murdered by the KKK in 1933 for having a Jewish law partner. He was a thirty-second degree Mason. In doing some research on the Masons she learned that the origins of the KKK were cemented to the Scottish somehow. She surmised that her grandfather had been murdered by his own people.

Jean was clearly guilty of moving a black couple into her all white neighborhood, and now the widow was living with her. As they say in theater, "If you are going to blow it, you might as well blow it really big!" That was one of her favorite expressions when she and her white horse were riding their adrenaline high together. She wrote down her experience because she didn't want it to be lost to the obscuring oblivion of time, and if Jean were to be called into a court of law her story would be available to refresh her aging memory, and so the story unfolded.

## TWO DEAD COONS BY JEAN STINSON LLOYD

One was a young black man. The other a furry little creature, who wore a black mask. The latter was splayed out on the main road below my daughter's house overlooking a verdant, urban creek corridor. The furry little coon was leaving nature's habitat when its entrails were shot out it's rectum ten feet upon fatal impact. The first dead coon, however, was my dear friend Cedric who had become like the son I'd never had, and his wife like a daughter. It was a clear, crisp morning in Asheville, NC. I was out for my morning walk and a Cuppa Joe. I couldn't help my sarcastic reaction when I saw the furry dead coon in the street. I knew I had become disenchanted with the American way.

I passed by the splayed fur and blood on route to my favorite caffeinated urban adventure in a town that I had visited at least once a year for the past four years, to log in another mother-daughter visit with my own flesh and blood. When I came back around, Rocky Raccoon was still in place. When we went out for breakfast that morning at Biscuit Head, Rocky was still in place. When we came back, guess what? You got it. No one cared. The furry little creature was just another casualty of a fast, mean, and nasty American landscape.

Cedric and Mari were an like answer to prayer. Admittedly, for the past forty-seven years I have been mildly obsessed with the post-slavery scenario. I was born in the San Francisco Bay Area in 1955, and in 1971 my parents sent me to visit relatives in Florida. My first cousin, a big strapping good-looking white male, twenty years my senior, was employed by the state's law enforcement and corrections industry. He asked me what I liked to do. I answered, "I like baseball." and that Vida Blue was my latest heart throb. Instantly I became a nigger-lover. I was really too young to process it fully at the time. I did not take offense with his roaring laughter at me, his West-Coast cousin, obviously being raised poorly out there in California; the land of fruits and nuts.

I went home at the end of that summer profoundly changed. I had first- hand knowledge, at the young age of sixteen, of the white way of doing business. I knew in my soul that hating a skin color was wrong. I shared this with my best friend. Her heart throb was Reggie Jackson. Together we began to enjoy black American culture because we could. Our parents let us skip school to drive my 1962 Metropolitan to the Oakland Athletic's World Series victory parade. There were not many white folks that day, and a whole lot of blacks. I was aware of this, but in the innocence of youth it just seemed natural. Nobody cared about color. We were all just celebrating our team's victory together.

The Bay Area was not like the South, and our parents had never told us we weren't allowed to fraternize. They gave us loving space to enjoy the world around us free of hate and restrictions. I guess my dad probably did say in passing once that marrying a person outside of your color brings a whole passel of problems, and would become a tough world to navigate, but I had no idea he was talking to me. In retrospect he was probably saying, "Don't bring home a black man." It was not out of hatred, but more from knowing how difficult the world can be, so why make things harder on yourself if you don't have to? My friend and I were content with seeing the Temptations at the Circle Star Theater, and listening to KDIA Lucky Thirteen radio out of East Palo Alto. We were actively enjoying a different culture in the autumn of 1972.

The opportunity of tasting the forbidden fruit never availed itself. Insulated instead, I lived the lily-white-world-on-the-West-Coast-white- privilege-lifestyle. I could never, however, shake the feeling of the life altering event at age sixteen as if it had been a gift from God. I'd had a profound spiritual experience that I was not willing to discard. There was a drive inside requiring me to speak out and tell my story to the world. If we ended up killing the planet, and everyone going extinct, then we wouldn't have a chance to clean up our mess and fix ourselves or evolve. This would seem basic to some and useless to others. Clean it up and be kind. Why not?

Through the next fifteen years I graduated from college, had a first marriage, and a re-location up to Reno, Nevada, elegantly re-named by the locals as "The Mississippi of the West." I never figured out who re-named the town. Was it the blacks or the whites? I would live there for many years before someone espoused that colorful moniker. I had no idea I was settling down to raise a family in a very white world. White was ubiquitous. The profound experience of my teen years had festered quietly in my system until our three-and-a-half-year-old-daughter, her dad Willie, and I all traveled back down to California to visit friends for the weekend. We were on our way home when we stopped into Spenger's in Berkeley for an early-bird special on our way back up the Sierras.

Our little family and a large, statuesque black woman in full ethnic garb were the only ones seated at the rear of the dining room. While the three of us were perusing the menu my daughter said, "Mama, look at that woman!" It's important to note that my child came out of the womb with good manners from the start so she knew better not to point and stare. I replied, " Yes, Darling," glancing over, "That's a black woman. Isn't she beautiful?"

Nothing more was said for the next few minutes, but occasionally I would glance at my little girl secretly studying our dining room neighbor. We placed our order with the waiter, the menus disappeared, and my daughter said to me, "Mama, that black woman is beautiful." I knew right then and there that if I were to die that night my job on planet Earth was complete. I had taught my child not to hate because of skin color being different than her own.

Years would go by in my insular white world. After my third divorce I ended up in a relationship with an older Italian man who had grown up poor in southeastern Ohio, born in 1939. He would boast that his county, forty-two miles north of the Mason-Dixon line, just across the Ohio River from Moundsville, West Virginia, was the only county in America without blacks. This was back in the fifties when he was in high school. At that point in my life I was desperate for companionship, and he offered adventure, exercise, and two young grandchildren to fuss over. It was a way out of my own family that offered very little love. My dad had passed away by this time, my daughter had gone off into the world, and my mom was my caregiving responsibility. I needed a pal. My other friends could see from the start that this guy was a mis-match, but I have always had the proclivity to ride into a difficult situation just for the purpose of understanding the other side.

Luigi's stories were remarkable. He was a classic little old Italian man who could relate to blacks because his kind had received similar prejudicial treatment way back in the day. He, however, made it clear to me that blacks were niggers. He had ascended out of poverty, made a couple of fortunes as a professional gambler, and was able to live the good life even though he had been investigated by the IRS for tax evasion. When I went with my biological daughter down to the Bay Area for the week between Christmas and New Year to help her move into another living situation after college, she dropped me off at the bus station in East Oakland at eight o'clock on New Year's eve for my return trip home. I was to catch the two o'clock in the morning bus. I had six hours to kill. No one else was there except for the two black male employees who ran the bus station.

Since there was nobody else in the station at the time, I sauntered over to the desk to ask a question. This then began my chatting with Elijah. Because we were close to the same age, it was easy to compare notes as we had both grown up on the same bay. He was a tall, strikingly good-looking man, nicely built and beautifully spoken, but only on one side of his face. The other side looked like he had survived a fire. I learned that Elijah's day job was head of security for BART. He was putting his four kids through college, and 'work' was this man's middle name. For a few hours my life was put on hold, I was able to dance with Jesus. I fell in love, not for romance, but for spiritual kinship. He was a person I admired greatly. I felt psychologically safe in his presence. We ended up at the front door of the station running security checks on the incoming travelers. I kept silent while he did his work as the customers passed between us.

I jumped on the bus at two in the morning after bidding Elijah a fond farewell. We pulled into Reno at six-thirty AM, and I was met at the station by my hot Italian lover. I told him about my experience. He went BALLISTIC. How could I possibly associate with a nigger? I couldn't help but laugh. I thought he was kidding at first, but no, he wasn't kidding at all.

I enjoyed my bus ride so much that the following summer, when my daughter asked me to come down for another visit, I said, "Sure. I'll take the bus!" I arrived about five o'clock on a perfect California afternoon. Oakland was a bustle. I remembered my friend Elijah. On my way through the station I figured I might as well check the front desk to say a quick hello. Sure enough, he was there. I said, "Elijah, it's Jean from Reno, I just wanted say hello."

He remembered me, we shook hands, and off I went outside to the curb for pick up. That was in the spring of 2009.

On June 17, 2015, another profound spiritual experience came knocking at my door. Twenty-five-year old Dylann Roof entered the African Methodist Episcopal Church in Charlottesville, South Carolina. He was invited to participate in a prayer meeting attended by the black parishioners. He then opened fire and nine were dead. It took the surviving church-goers two hours to forgive him! I was still caregiver to my mother at that point and was able to station myself in front of the TV. Back then, atrocities were still commanding adequate coverage. I would retell that story to my white peers at least a hundred times before I would have the privilege of meeting Mari and Cedric.

It was 2017 when I first met them. Willie, my ex, was the ranch manager for a mutual friend of ours up north, in the stark location of nowhere. The Old Beaver Ranch was a place where Burning Man people could stage their participation before entering the event. The Burner's support facility sported an off-the-grid lifestyle which ended up killing Willie in the end. People could get as high as a kite or as low as a mole, there was no law enforcement to interrupt the party. Willie would nurse a 24 pack of PBR most days, and that's how Mari and Cedric ended up with Willie's job after he died.

The ranch owner, our friend RD's mother, was the first person Willie and had I met when we moved up to Reno in 1984. You could say we all grew up together. Our kids, our dogs, our horses and our lives. The lily-white-world of northern Nevada was a well kept secret of snow-clad peaks and expansive valleys, mimicking a poetic metaphor of life. With Willie's passing came a second-course helping of my old flame RD, who'd had a crush on me since our 20's. He was the first born son of the ranch owner and Willie's boss before he passed away. We became lovers again probably because of Willie's passing. There was comfort in old habits.

Carmine B. Littleworth

The autumn before he died I took Willie and myself back to North Carolina to be with our daughter for a two week vacation. He got sick on the way home, and never could shake the bug. He was living in a cargo container that had been re-purposed into living quarters out on the ranch in snow country. It was heated with a five gallon tank of propane gas, the same as a BBQ grill. It would have been fine for a well person, but that particular winter was a real hard one. Willie had too much pride to ask for help. He suffered from a bacterial infection on his transplanted heart valve when he went down. He died on tax day 2017.

RD, who was a half Jew/half Christian, and I were driving south one late spring day when he first called Mari and Cedric for a phone interview. Hiring a black couple was still fairly avant guard for Northern Nevada back then. I kept asking him if his mother would accept black people. His reply was, "The color of your skin shouldn't matter!"

Because I was enjoying the relationship with RD up north I would occasionally drive the two and a half hours for a visit to help out with whatever needed doing on the ranch. The week before Burning Man was insane with mostly self-centered spoiled white people, young and old, doing their version of what mattered only to them. I teamed up with Mari and Cedric to help these silly fools, who couldn't pour piss out of a boot, stage their outrageous props. Cedric was a military man with well-trained army skills that had been underwritten by the United States tax-payers. Mari was his bride of fifteen years. They had met when she was fourteen, and he sixteen. They had never known anyone but each other, I surmised. They were fully dependent upon one another. This was unique in my circle of reference as a three time white divorcee.

I was up at the ranch helping out about a week or so before the event opened. It was too much for me. I don't know what it was that caused me to spend the whole day sobbing, but something deep within was triggered. It may have been all the nasty, selfish people that the young black ranch managers and I were helping. It may have been Willie's death as he was the father of our child. It could have been a need to reconcile my spiritual PTSD from post slavery Jim Crow. I don't know. But I spent that whole day doing laundry, cooking and sobbing to Mari my whole life story. She was like a mother to me! I had waited my whole life to tell my story to a black person, and to relieve the sorrow of what my color had done to their color. I had the same experience with the local Indian tribe, the first nation, because my ancestors had been Indian and buffalo killers.

By the time 'The Man' burned to the ground the young couple had racked up a few months of dedicated service to the ranch. It was time for RD to give them a little time off for good behavior, except that he wouldn't. We tried to convince him to let the young folks come down to the urban area, stay with me, enjoy themselves for an evening, pick up supplies, and then head back up to the ranch on the following day. It would have been a safe, occasional, twenty-four-hour turn around.

RD refused to let them leave the ranch. If someone were going into town it was going to be RD, and only RD. Later on, after Cedric died, Mari would recall how RD would want to be the one to purchase her necessary feminine products at the store. He had promised to provide them with a car, but the car never came. So they left him, and they turned up on my front door step saying, "Remember how you said if we ever needed help, just ask?" I said, "Yeah." The next thing I knew we were all living together. I should back up and add that the three of us had been having many long conversations about everything from soup to nuts. We had an instant affinity for one another. I needed the black friends I had always wanted, to be able to heal my broken heart over Jim Crow, and they needed a safe place to land. We all lived together for quite a few months, which gave them a chance, rent free, to figure out their next move. We shared something important: Faith in God. I had been saying to them for a while, "I did not access Jesus Christ through white people. I met Jesus through black people." Those old negro spirituals were songs that spoke to my heart. Admittedly, I wanted to fix the world.

I did not want them, however, to become permanent fixtures in my household. I was over-involved with a building project at church which took all of my time and strength away from my own projects of house remodeling. I could not have others underfoot while suffering through a kitchen with a broken sink, no countertops, only one bathroom, and dog-soiled carpet throughout. I have always kept a cold house at fifty degrees. I saw no need to make things more comfortable for the young folks because I needed them to be back out into the world ASAP.

The time we had together was a hundred percent pleasant, a hundred percent of the time. I never once heard a cross word or a disrespectful comment. They never complained. I let them use my bikes to get down to the convenience store, or they went with me into town to shop. They purchased their own food. I definitely created an environment for them to be motivated to be on their way. If they had been a young white couple, I probably would not have offered a place to stay from the get go. There was something very safe about this pair.

It was in the dead of winter when I witnessed Cedric having a PTSD attack. The three of us were on the north side of the house discussing our plan for the day when we all heard the explosion of the avalanche cannon going off up on the snow-covered mountain overlooking the neighborhood. I saw Cedric's big white eyes, just like in the cartoons, get huge and then freeze. I put my hand on his shoulder and said, "No problem, that's just the avalanche cannon up on the mountain." He relaxed. We all laughed and continued our day. I knew he had PTSD from the war because Mari told me so, but other than that I never saw another episode. Cedric was a two-time-tour-of-duty in Afghanistan war veteran. I remember telling him he should wear his veteran's baseball cap.

I think they may have had hopes that I would want to take them on as mini ranch caretakers, which would have been fine had the house been already remodeled. I could have left town and had no worries just like the arrangement they had up north at the Beaver Ranch, but on a smaller scale. I introduced them to a neighbor who was a very kind and warm Korean War Vet, but the neighbor above me, across the street, did not speak to me for over a year after they moved in, and a bit longer after they moved out. I got the negative vibe from the wealthy white woman, across the street, on the second day after their arrival with their trailer-load of belongings. I refused to let my new houseguests do anything to help out inside the house, or outside, because I was fearful that someone in our neighborhood would be making disparaging comments, and the thought did cross my mind to fear for all of our safety.

In spring the following year, they announced that Mari would be actively pursuing a career at the local hospital. She had prior experience and skills in the field. The interview process would require trips to town. I chose not to loan out one of my five cars because of being too nervous about the restrictive systems of police, hearing so much about driving while black. I did, however, volunteer to be their driver. I'm a mom, so it seemed natural to have kids once again to drive to all the various requirements.

Cedric and I would sit in the car, sometimes for two or three hours, waiting for Mari. We would discuss politics, religion, bad people and good people and we always had their dog Prince along for the ride. He was an unofficial service dog to both of them, and I am an ardent dog lover. It was a match made in heaven. In a way, God dropped off an instant family right in my lap. I was enjoying the relationship with the three of them very much. They were first class American citizens, and deserved a leg up. You could say I did it for my country but, really, I just enjoyed being in the company of people with such good manners and intelligence.

Once Mari secured her position at the hospital in Reno we drove around looking for an apartment. They found one a half block from the Truckee River and a block from the bus stop. We spent a day moving their stuff into town. They finally had their own place. I gave them my old electric turkey roaster because the only cooking option the apartment offered was a two burner built- in hot plate. I took them to the grocery store and bought them a shopping cart full of groceries to stock their larder, just to get them started, and to wish them well. We would get together every few weeks to go out for a meal, and keep current. It was my way of keeping tabs on these young people who had become my kids. I wanted them to succeed.

I think Mari had a harder time accepting the person-of-color back seat position. Her lighter skin, half Mexican/half black was a beautiful shade of Caramel. When all dressed up she was stunning. Mari exhibited a regal presence and conducted herself with flawless demeanor. Cedric was very dark skinned with dark eyes and a big engaging white smile when he was comfortable with the situation. It seemed like she was less able to tolerate the unequal treatment dished up to people of color. She said she had come from a very proud military family where success was measured by hard work and honesty, not just by a privileged dermis. I had a taste of what it must be like to be black when we were up at Tahoe shopping for the afternoon in the fancy stores at Stateline just over the California border. Prince, the dog, was along as usual. Cedric and I would take turns hanging out with the friendly Chihuahua, while Mari enjoyed herself in the stores. It was Labor Day weekend, the shops were packed, and the atmosphere festive. We were all hanging out, walking along and shopping. Mari was patronizing a store where she found Christmas presents for their family. When it came time to pay the bill, which came to eighty-nine dollars and a bit of change, the clerk stated she could not break Mari's hundred-dollar bill. I can't remember how Mari handled it, but we ended up having a discussion afterwords about how this type of treatment was the norm. I pondered on this for a long time. The clerk couldn't pull a ten dollar bill and a few pennies out of the till on the afternoon of the busiest shopping weekend of the year!

Mari's work at the hospital continued through to the next spring. She did not ride the bus. Every day, Cedric and Prince would walk her the mile-and-a- half to work, and then walk home. In the afternoon he would walk back to the hospital to pick her up. She was not, however, comfortable with her employment arrangement in Reno and longed to be back on the East Coast, for some reason I never understood. She landed them a new position in Charlottesville, North Carolina. They had to start in three weeks as resident caretakers of an elaborate old-folks community. This was one of the many times they would use me as a reference to secure a new job.

They needed to buy a car. They searched around online and I volunteered to drive them around to the various car dealers. They looked at used cars. They wanted to be able to pull a trailer. Cedric did all of the driving. Mari would frequently suffer from panic attacks behind the wheel. Cedric's military background made him the person both Mari and I counted on to be the man, and he was. He was always calm and focused on the task. Never once did I see him waffle from any job needing to be done. Mari pretty much ran the show, and we all took our orders from her after she listened to input from her husband, and me too, if I were in the mix at the time.

We arrived at the Jeep dealership shortly after viewing a huge, used, five year old black SUV Cadillac with dark tinted windows at a cut-rate car lot. They took it out for a test drive. It was nice, comfortable and roomy. It must have been a 2010, a real gas guzzler, but it was not the car they wanted. At the dealership they found just the right car. It was electric blue and brand new. We were there until well after closing, finishing up the paperwork. They had no existing car insurance policy. It appeared, from my outsider's position, they had clean credit, but not a lot. We were there for at least two hours after closing. Cedric's military service must have helped them quite a bit. We were to return early the next morning to take delivery. I say we, but it was them. I was just their driver and shepherd. I got to be a mom again. I lived twenty miles south of town so a fair amount of driving was needed to pull off our project, but I was enjoying helping the young African-American couple be yuppies in the white world.

The next morning we went out to breakfast, dotted our I's and crossed our T's, and stopped off at the U-Haul store to make sure the new Jeep would accommodate a trailer hitch before Mari and Cedric were to sign on the dotted line and take delivery of their new vehicle. We made an appointment for later that day to purchase and install the trailer hitch. We went back to the Jeep dealer to pick up the new car. I took a cell phone photo with Cedric driving, Mari riding shotgun, and Prince on high alert in the front seat with Mari's arms wrapped tightly around the family dog. Cedric leaned forward and smiled for the photo. I could tell he was a happy family man excited about his future. It was probably my fondest memory of all. I felt like I was serving my country. These were good young people capable of going out into the world to do good things. They were always devoted to helping others, and willing to go where the Lord saw fit to take them. I was happy to help. Now looking back on everything I wish they had bought an old clunker that would never have made it out of town.

But they did make it out of town. I treated them to the trailer hitch and a trailer rental, and the next morning we were all at their apartment ready to load up. I stayed at the curb guarding the load while they brought everything down in the elevator. Cedric being a military-trained packer, a natural born body-builder, and Mari classically female, got the job done in an orderly fashion. Together with their dog, they were the cutest couple. She was a little bit of a fire cracker, and he was always the perfect gentleman. Both were excited to move forward. I was so happy to be able to help. Being in their company felt magical. They never asked for anything. They were like my own children. That's why they called me Mama Jeannie.

That afternoon they got on the road and started driving east. We would check in with texting at the end of every day so I could track their journey. They let me know when they made it to North Carolina. They were on their way in the world, and I was their helper. It really was like a dream come true to be able to have black relatives. It was that old heal Jim Crow syndrome, and I was loving it. I even came up with an idea so that every American could help heal our inter-racial divide. White people could sign up on Facebook to be adopted by a black family. They could call it "Adopt-A-Honkey!"

After they got to the East Coast, Cedric and Mari kept on with the practice of switching jobs regularly. The Jeep sure was paying off with its new trailer hitch. I would be their reference. The new employer would call me, and I would sing their praises. Then they would move on to the next place of employment. At some point they switched out of the old-folks-home industry and into Air B&B management. That is how they ended up in Boston. I had been to Boston before, knew the area, stayed with a friend, and enjoyed walking to the Arnold Arboretum as many mornings as possible. Those big fancy houses of Southwest Boston had given me a false sense of security from the start.

The next spring, March of 2019, John and I drove a load of Willie's personal effects back to my real daughter in North Carolina. That was just a few months before Cedric's murder. I met the black preacher next-door to my daughter and son-in-law's newly purchased first home which was next-door to a black church. The preacher was willing to chat that Friday morning when we first pulled in with the van and trailer rig from the other side of America. There was no way this man-of-the-cloth could've known my lifelong spiritual journey through America's race issues. I tried to explain my position, and expected him to welcome me to church that coming Sunday. After all, I was not a racist, but rather a sister in Jesus who wanted desperately to heal America's mistakes. He listened and we had a good conversation, but in the end he informed me, "We separate." I fully understood immediately, and moreover, felt privileged to be turned down by a black person because that meant that I was participating in the healing, and not just asleep or deaf to the national mayhem. After all, it was about time black people were able to say no to whites. This was just another signal from God that I was tugging on the thread of healing our nation's woes.

On that Sunday afternoon in September, 2019 I received a phone call from a number I did not recognize. I don't answer a call if I don't recognize the number. The same number called again, and still I did not pick up. Five minutes later I got a text from Mari saying, "Answer the phone!" <It was a matter of life and death!.> I did. She told me Cedric had been killed by the Boston PD!

I now know how a mother of a black man in America must feel, living in constant fear that your child will be killed by the police. I had already been dreading this for about a year prior to Cedric's death. It's not really possible for any mother of a black son to feel any different. I was the first person Mari called. She was too afraid to call her family or his. She called Mama Jeannie.

After the initial shock, I had no other alternative but to become a warrior. Wrong is wrong and right is right, and never the twain shall meet, as the saying goes. All my years of watching black people videos and learning about what it must be like to be black in America would finally have a purpose. I would avenge my son's killer. Groping at straws was my first gut reaction. I Googled and called as many helpful avenues as I could possibly think of, from Black Lives Matter to Bryan Stevenson. There was not a lot of help. I tried to drop a few seeds on social media, and leak a small drop to the press, but not aggressively. I was too afraid, but of what? The KKK? It was up to Mari to call all the shots. We texted volumes.

Somewhere along the line the young couple had picked up an extra dog, William, a much smaller Chihuahua who looked up to Prince for direction. Someone, some entity, paid for Mari and her dogs to be put up in a nice hotel after the shooting. The police arrived at the hotel to try to get her to give her statement. The panic attacks disabled her from speaking to them. Her very best everything in the world had just been murdered. Why would she want to give the police information that would help them cover up their own crime? If Cedric would have been a person of questionable character it would've been easier for me to shrug it off. If Cedric had been anyone but an exceptional human being I probably would not have been so motivated to bring change to America. Cedric, however, was exceptional as a man, soldier, husband and friend. For all intents and purposes Cedric was a rare soul never to be heard from again. He was silenced forever at age thirty-seven. Were the police now in the practice of murdering our veterans? Maybe Cedric had too much knowledge from the war? After all, he had worked in intelligence. Maybe the government needed him to disappear from his quiet and modest life with the occasional mild PTSD event? Mari wasn't speaking. Mari refused to speak.

Carmine B. Littleworth

A couple of days later someone moved Mari and the dogs to a cheaper hotel. Mari wanted me to come back to the East Coast right away but I couldn't. I am no longer a spring chicken and every time I travel I get sick. I had already booked my flights for a mother/real-daughter week at the end of October, and the idea of making two separate trips East in one month was daunting. I discussed this with my blood-daughter. She stated she did not want me to engage in something that would shorten my life as she had just lost her dad a few months earlier. I appreciated that she cared. It gave me just enough space to be able to tell Mari I would arrive the last week of October. In the meantime I let her know, "We would get through this together." All I could do was to be supportive from long distance.

Just before Cedric died he and Mari had secured another new job. This time they were to be the new inn managers at a B&B up in mid-state New Hampshire. Again I was their reference, and again they got the job. It had nothing to do with me and everything to do with Mari's ability to move her family into the next gig. I was happy I could speak so highly of the wonderful people I dearly loved. They were to move up north in two days, but Cedric was taken out by the police instead. Would it have been a different outcome had my son been white? Mari was still not talking to the authorities. Would this case just fade away into the abyss like so many others?

I did not hear her side of the story until we were face-to-face in New Hampshire. I think black people are naturally very guarded and don't give out information easily under any circumstance let alone the murder of a dearly devoted husband. In the meantime the Suffolk County District Attorney had put the crime scene under investigation. I was able to follow the news stories over the internet. There were at least five news sources following the story including the Boston Globe when it first broke. I only had to Google Police Shooting September 29th and the latest news reports would appear. In the meantime, John and I had made plans to visit my dying uncle up in Oregon, so we set out on a road trip going north via the redwoods of the Pacific coast.

We were just south of the Oregon border, on Highway 101, when I got the call from the mortician to pay the bill for Cedric's cremation. I volunteered to pay because I did not want any questions from anyone, such as the Boston PD, about whether or not Cedric had people. I had been through this routine before with other family members and I needed for Cedric to be cared for even in death. I needed Mari to have one less struggle. After getting home from Oregon I was back for about a week and then off on a red-eye to Boston.

I arrived at Logan around midnight, got a rental car and started driving north. The lack of sleep was easily manageable when I thought about Mari's loss. I chose to fly into Boston in the middle of the night because I did not want to deal with a town filled with so much hate for black people. I just wanted to get in and get out, which is exactly what I did. I plugged North Conway, New Hampshire into my phone and off I went into the dark and isolated back roads of New England.

It turned out all right; I arrived safely at four-thirty in the morning. The route was a dark maze of country roads straight out of the Twilight Zone for this old Westerner, who was used to long straight stretches. It must have been around three-thirty when I finally reached a little all-night market. I chatted with a young disgruntled male employee for a good ten minutes or so. He wanted to know my story, and I asked him about his. He was a white Alaskan native transplanted to No-Wheres-Ville, New Hampshire, stuck in a low paying job, on the night shift. He encouraged me not to purchase from the store because the owners were rotten. All I needed was a bathroom so I graciously accepted his candor and got back on the road.

Carmine B. Littleworth

Not wanting to disturb Mari's sleep so early in the morning, I found my destination, parked across the street, reclined my seat back, and took a well- deserved snooze even though I was fearing the police. I fully expected to hear a knock on the car window and to be woken up with a blaring flashlight. I knew to keep my hands on the steering wheel at all times until given permission to search for ID, or to open the window, or to step outside of the vehicle. Growing up white I never feared police before. I must have logged at least a good solid hour of rest. It was six o'clock. I knew she started work at the B&B at seven fifteen. I needed a cup of coffee just to feel human. Slowly I eased my way through the small tourist hamlet to an all night market with surprisingly delicious coffee. I texted Mari at six-thirty. She welcomed me into the living quarters that she and Cedric had expected to enjoy together.

The Arbuckle Inn had been serving travelers since 1863. Cedric and Mari would finally have landed the job of their dreams. The location and setting were of fairy tale quality and the owners were the proud parents of two adopted black children. Cedric would have been an asset to the historical resort town. Mari felt they got the job, in the first place, so the kids would have more similar faces around. Cedric didn't pay attention to color. He was good with every one. Back at the B&B in Boston the people in the neighborhood were reporting on the news their sadness regarding the loss of a lovely soul who had walked and picked up after his dogs, and would always chat with neighbors.

Mari went to work that morning while I crawled into a perfectly made bunk bed. The new apartment had been the initial living quarters of the owners of the Inn. Now they were living off site. Mari was working the front desk and living in the apartment. Cedric would have loved it and fit in beautifully! She returned to the apartment at eleven. Because she had to be back at the front desk by three to work her afternoon shift we decided to walk the dogs and get a decent meal at a local cafe. At a nearby park we leashed the pets and set out on foot around the lake with all the glory of a brilliant fall display. Mari began to tell her side of the story.

At first I didn't think I could handle it. She told me it would help her if I listened. I agreed. Our walk was about one hour. It was a loop around a large quiet pond below a grand rock monolith, a toe of New Hampshire's highest peak, Mount Washington. It was $8.00 to park and walk your dogs with Mari being well equipped with her leash-mounted doo doo retrieval system. This may seem like an unimportant detail, but after she told her story, after the walk around the pond, which they called a lake, and after we dutifully harvested our dog crap from the leaf carpeted forest floor, I asked for directions to the trash can. The white male attendant, middle aged with a thick New England accent informed me the park was a pack-it-out park. I resounded my approval of their intelligent system, but secretly wondered if he had hidden the trash can from us because of Mari's color so we would never return again. I even donated to the park the remaining two dollars out of the ten dollar bill just to drive home the point of our cooperative spirit!

As Mari was unfolding her story on that gray but mild mid-day walk-a- bout, the pieces of the puzzle began to make sense. Because I was the first person she called after the shooting I was the one to get the first layer of the story. Now I was receiving the next layer of the story. On the evening of September twenty-eighth she and Cedric were having a farewell visit with their newly made friends. They were to leave Boston two days later to go up to their new place of employment in New Hampshire. The male friend may have lit a joint, and that could have been what set off Cedric's PTSD. Both Cedric and Mari were non-users but did not judge others. Cedric set off on foot, on the sidewalk, walking away from the B&B. Mari was right behind him, comforting him, like she always did. She was a well trained military wife. They were engaged in, probably, a somewhat loud conversation but not screaming at each other. The time was 10 PM. Something triggered a neighbor to phone in a domestic disturbance call. Law enforcement arrived and began talking to the young couple. At some point Cedric and Mari became separated by the cops, which was the first fatal mistake because Mari had been the one managing her husband's PTSD for years. Soon the SWAT team arrived. For some unknown reason Cedric retreated back toward their apartment, and up into the stairwell to the digitally locked apartment door. Mari was able to make it back in time to listen to law enforcement speaking to her husband. Cedric was telling the officer he had a gun behind the locked door. I got the feeling from Mari that the officer had encouraged Cedric to unlock the door and retrieve the weapon. Mari heard her husband tell the officer, "Look at all I have done for you, for our country, and look at how you treat me." The next thing, as reported by law enforcement, Cedric brandished a weapon out of the smashed out second story window that he had broken out right before one of the officers on the ground took him out with a fatal shot to the chest. He was then allowed to bleed out and die instead of the police calling for an ambulance.

The police refused to tell Cedric's widow the truth. They lied. They told her he was getting medical attention. Is it the new American way for our police to kill our vets? Should we all consider law enforcement to be armed and dangerous? I certainly do. Do we need to implement a system in communities where PTSD vets can alert neighbors of their conditions that develop as a result of serving our country? Can neighbors help look out for our vets? May we, please, make this law and call it Cedric's Law? They say ask for what you want. I have.

# Ninety Three Percent

After Jean finished writing down her testimony for Cedric she felt better. She had written it in case she got called in as a character witness to a court of law. She was a lot like her emotionally challenged mother. If there was an injustice she would fight. She would not let go until the job was done. Her frustration with Mari being unable to act added a certain extra monkey wrench to the project, but Jean stuck with it because her father had taught her to be a person of principle. Her goal was to make Cedric's life matter. It must have been that Mari was just a very chill person. It appeared she did not have the nature to seek revenge and pursue justice. The two women had grown very close. There was less than two weeks left before Abdul would arrive in America to claim his bride-to-be.

The plan for him to come to the states to meet Mari's family, including Jean and John, unfolded day by day. The new couple would then drive the Jeep back to the East Coast where they would then become residents of NYC. After Mari announced that Abdul would be coming to America instead of her going to Turkey, a large shift occurred with the two women. A certain trust and peacefulness developed because of less stress and worry. Jean could now think of Mari as the "Black Cinderella" where dreams really do come true. Mari began going next door to socialize. They went back to taking adventures into town, going on walks, museums, etcetera. They both knew their time together would soon be ending.

With two weeks remaining before the big day, when Abdul was to arrive, they set off one afternoon to visit the newly opened First Nation Tribal Museum which Jean had seen featured on the local news channel. It was the reservation where she had taken her blood daughter to church as a child. Jean was a natural born church-goer. She and her daughter attended as many as ten churches or more while growing up. Church hopping was a regular practice for the mother- daughter pair. It was hard to say if she was after spirituality or just nosey about how people interacted with each other in the church setting. They ended up quite involved at the Baptist church on the reservation. She and her daughter participated in the Christmas pageant one year. The hymns had been translated into the Native language. There were other white people in the pageant also, but Jesus was indigenous. When the preacher's wife gave birth to their second child Jean stayed over night with their first daughter, who had the same name as Jean's daughter, a name that reflected an element of nature similar to the sun, moon and stars.

Arriving at the new museum on that gorgeous autumn afternoon, Mari and Jean were welcomed by an indigenous woman. She asked if either Jean or Mari had been to the newly designated National Register of Historic Places museum and grounds. Jean related her church memory to the woman with a tear in her eye. Then she gracefully indicated she needed to begin seeing the exhibits in order to regain her composure. The museum woman immediately latched on to Mari. Jean felt comfort hearing the two young women connecting. She was enjoying the background chatter of the Native woman's attention towards Mari. It gave Jean a chance to relax and immerse herself in the cultural energy that she dearly loved. There were old photos of tribal people before they had become Americanized, and after they had become Americanized. The purpose of the museum, the docent had stated, was to tell the story of the journey of the original-nation people after the arrival of the white settlers.

Jean had not realized she was walking into an emotionally charged atmosphere. She was not aware in the slightest as she was lost in thoughts of her own past. She was relaxing into her old memories at the back part of the museum. Mari found her at the reader board that talked about the oppressive white culture as it pertained to the linking of the first nation people with black slavery.

Finally after four hundred years the people of color were realizing they all had something in common. With only seven percent of the world having white skin, and the balance being BIPOC—Black and Indigenous People of Color— the math said it all. Ninety-three percent BIPOC, against seven percent white, was an impressive ratio to Jean. She told Mari the numbers because she wanted her adopted daughter to feel her power in the world, even though she was compromising her own color in the process.

Jean was devoted to helping out the black people because she was sick to death of America's white privilege abusing its power. Most white people were in denial according to old Jean. She had experienced the ugliness herself, first- hand, even though she herself was white privilege.

Back in the eighties and nineties, when she ran her landscape company, she hired nothing but illegal Mexicans and drove around in a beat-up old pickup truck with no air conditioning. Five days a week she was physically exhausted and covered in dirt head to toe.

She'd be driving her men back to their ghetto-style apartment at the end of the day waiting at a red light next to a giant Lincoln Navigator. Jean felt like two cents waiting for change next to the vehicle with all the windows rolled up driven by an itty bitty blonde trophy wife sitting behind the steering wheel higher than the woman's path of vision. She wondered, at the time, why she hadn't sought out a white collar- man with big bucks so that she too could've been a trophy wife. But she was stuck in her dirty world that had spiritual benefits instead. She had the privilege of seeing how the other half lived and being comfortable with people who weren't the same. She wasn't very comfortable socializing with the nouveau riche white people, who were completely convinced that their feces had no particular odor.

Many of her landscape and garden clients were from old American money. Those people were decent. They had hearts. They were kind. But that new breed of white money that came on hard and fast in the last quarter of the twentieth century was clueless and unable to have any awareness of the bigger picture. Jean called her color of peers Low Consciousness. From her perspective it appeared that as long as their lavish needs were being met, then that was all that mattered. If people were suffering, and the planet was dying, there was nothing that the Low Consciousness Folks wanted to change about it. It was not anything that concerned their daily routines. They could continue to sit in their fancy A/C cars, at drive-through coffee windows, and not be bothered with the hell that was brewing alongside their favorite latte. The lack of compassion and awareness for others was profound.

Jean followed the United Nations Sustainable Development Goals. She had a sneak preview of what was to come long before many of her fellow boomers. Her involvement started on the Linked-In social media platform back in the early days. Somehow she connected to a group of men from all over the world who discussed the problems that were plaguing life on Earth. The players in the group, day after day for a couple of years, would shoot their wads into the virtual cloud to discuss what was wrong, what needed fixing, and would make possible suggestions on how to avoid a dead planet and the extinction of life on Earth.

There were scientists, economists, environmentalists and professional people like Jean who had observed the changes to the environment over time, and knew that entire colony collapse was imminent. Water on Earth was Jean's big issue. She had lived her whole life in the arid West. She saw streams go from crystal clear and drinkable to algae ridden and undrinkable. The last time she drank from a stream in the Sierras she was pregnant. She got a water born disease called Giardia during her first trimester. That, along with the morning sickness, gave an interesting start to her day as she was on her way to work at the landscape construction site with no easy access to a toilet. A trophy wife she was not!

So passionate she was about water that during one of the extended periods of drought she decided to turn off the water to her master bathroom toilet. Her plan was to harvest the water from the tub spigot while waiting for it to warm up for her bath. Two gallons of precious water would escape down the drain while waiting for the hot water to arrive unless she caught it in a bucket. She told her science friends on Linked In about the new project. One Doctor of Science from India reported it was a dangerous mistake to turn off the water to a toilet for fear of coliform bacteria.

Around the same time as turning off the water she decided to convert her toilet to a squatty. She had visited in China and preferred the system of squatting, especially when she learned that Americans had weak legs and knees because they no longer squatted. Apparently, the invention of the toilet had turned westerners into weak people. This made a lot of sense to Jean. She tried to invent an alternative toilet seat from a large white kitchen cutting board that could be squatted upon for the purpose of elimination. Unfortunately, it was a total flop and terribly uncomfortable to perch high on top of a cutting board, with a hole in it to take a shit.

And of course being naturally obsessed with the study of waste management, and over half of the world having no access to sanitation, she decided to try using her hot soapy dish water for toilet flushing. She envisioned she was on to something really big!

Thinking she had made the greatest discovery ever, she was ready to reroute the kitchen plumbing directly to the toilet so that the toilet could receive a daily cleaning of hot soapy water. After a few days of collecting dish water in a bucket and delivering it to the toilet bowl she ended up with a toilet ceasing to function. The inside surface of the gooseneck, which is what prevents sewer gas smells from entering the house, became coated with adhering soap scum. Flushing was no longer an option. She spent three months trying to heal her toilet. All she got was frightening levels of un-flushable Poop Soup. She had successfully, and completely, destroyed her loo.

On a beautiful spring afternoon she ripped out her toilet with no replacement on the horizon. The item became garden art. She scrupulously cleaned her bathroom floor and the opening to the sewer line. She inverted a plastic yogurt container over the opening. Jean turned off the water to that toilet for seven years and never had a problem with deadly diseases.

This woman was obsessed with clean water on Earth! She refused to give up her science project. For all those years in that bathroom, which was her very own private space that no one else would ever think of entering, she created a make-shift squatty out of a black ABS shower drain insert and a half gallon hole-in-the-bottom yogurt container with lid. She was thrilled when the time came to clean her toilet. She'd just throw it all outside, give it a hose job and let the sun sterilize the alternative device.

The point of this long drawn out toilet story paints a picture of a woman who cared deeply about the planet. She definitely thought blacks were abused by whites, that was a no-brainer for Jean Lloyd. She also felt strongly that her fellow white people had abused and destroyed the planet with their technological advancements. They just couldn't leave well enough alone. Everything always had to be just a little bit better every day. Over time those so-called improvements ended up painting everyone into a corner with no escape. People who were used to their creature comforts didn't want to give up anything. Instead they wanted more of the same unnecessary crap. The problem with more is that every action creates a carbon footprint. If you drive a mile to a cool hiking destination instead of walking you've laid down a carbon footprint. If you take an airplane instead of a train, that's a big footprint. If you don't support the organic food industry, you are personally killing the planet with fertilizers and pesticides made by the oil companies that make those items as by-products of crude oil. Profit and dividends to shareholders run the show. In short we have all been duped into believing that the lives we are living are normal. This was Jean's obsession, and Jean, too, was also a shareholder. The difference was that Jean was willing to see the contradictory and heinous role she played in sucking the tit called capitalism.

Back at the museum and immersed in thought about the First Nation dwellers maintaining sustainability for over ten thousand years, she lamented her own ancestors who had worked so hard to settle the country only to have it destroyed by over-consumption of natural resources. The devastation that had occurred in the last fifty years— during Jean's lifetime—had tipped the balance for good. Obvious for some but not for all. She belabored the point to no end.

Mari and Jean were still in the back room of the museum enjoying their visit. Jean gave Mari an elbow nudge, "Hey, check this out. Here it is. This is how they lived." She pointed to a photo of a little mound-shaped hut that was slightly shorter than the native people standing in the foreground. Mari saw them as living in poverty. Jean saw them as people who knew how to live sustainably. Jean, always the theatrical opportunist, seized the moment to espouse her approval of the sustainable lifestyle dwelling verses the planet- killer buildings causing all the destruction. No wonder people of color hate white people. Whites are the destroyers, thought Jean. Every time she had a chance to drive home her point to save-the-planet she never wavered. Mari was always polite to her Mama Jeannie, but she was also very tired of hearing how everything was being destroyed.

She had woken up one morning a decade earlier with the realization of her divine purpose to heal the entire planet before she died. This came with an intense need to keep it a secret because she knew no one wanted to hear about it anyway. When it first came to her she kept it to herself. She thought it was ridiculous. How could one person heal the entire planet alone? That was the point. It would take all hands on deck.

Decades before, when she was hanging around with the Hindu community, she was told by the Indian Guru himself that she did not need to do the work; she only needed to show up. At the time she had no idea what it meant. But as time went on she learned that ego wanted to take credit for work, whereas spirit just let go and let God direct the outcome. She believed that God put in front of her whatever God wanted her to do. She discussed up her belief system, but never imposed it on anyone. She realized the Guru was talking to everyone through her. Everyone needs to show up. It's by the grace of God we live and the God factor can pull the plug on us all at any time. Jean contemplated existence yet again.

After the two women left the First Nation Museum, Mari told how the attendant had wanted to connect with her in a way that was slightly uncomfortable. Mari was being careful not to burn any bridges with Jean because she was her current lifeline who had offered her a place to stay and refinanced her car. Jean's white pigmentation was valuable to Mari. So when the Native American attendant appeared to be cultivating Mari, a fellow BIPOC, as one of theirs, it was an indication, to Jean, of trouble brewing for the white world. The BIPOC's were gathering their people one person at a time. They were beginning to realize the quantum physics of humanity being 93% POC. There was no way the white folks could stand a chance if all the people of color became united against the NoPOC's. The white people who believed in diversity might be left out in the cold. They'd be shunned by white and shunned by color. What an interesting predicament, thought Jean.

Mari told Jean she felt as though the young First Nation woman was trying to encourage a sisterhood status. Jean and Mari got back into the car and looked around at all of the old stone reservation buildings that had been abandoned and vacant for years. They were one of the old Indian schools from America's troubled past.

The old stone buildings had been built by the tribe under the supervision of the governing white population and operated as an Indian school until 1980. They were magnificently constructed from colorful rocks gathered locally. The photos inside the museum showed how in the beginning there were no smiles, but after many decades the natives had become sanitized into good looking healthy, smiling people. It looked as if the tribe was showing other tribal members how far they had come in the white world under precarious circumstances. This conflict was unsettling for Jean. She wanted the Earth fully healed before she was dead and felt that as long as human beings were continuing to engage in racism and prejudice because of skin color, there was no chance to do the important stuff like healing the planet. Jean felt more credit should be given to the old ways as the old ways had been sustainable for thousands of years and the new ways were evolutionary suicide. Jean's obsession raged on.

# CHAPTER NINE

# Black and White

It was a hell of a conundrum for Jean. White people made the mess and now everybody needed to get past their differences, fall in love with the Earth, and practice conservation of natural resources. What's the problem with that?

The world according to Jean stated it wouldn't be easy, but it could be done if everyone on Earth demanded reconciliation and restoration. Unfortunately for white privilege, the BIPOC sector of the population had woken up to the power of its numbers. As it stood, the white folks didn't stand a chance as an evolutionary entity. They were slitting their own throats by not being kind. Quantum physics and the snowball effect had finally caught up with the diminishing white faces and the projection wasn't looking good for the honkeys and the gringos.

Jean was pissed off at white supremacy. In the world of systemic racism and the KKK, she felt like she was being set up for a disaster. She saw the existence of whole systems and not just parts. When an imbalance of power presented itself to any living organism, causing great pain and misery, either animosity or extinction would follow.

Jean saw the many systems of nature clearly. It made her a misanthrope. She didn't choose it, she was stuck with it. She had a hard time fitting in because all of her white friends liked the way they lived, and didn't want to be thwarted with having to give up their perceived comforts. Extinction Happens. Jean liked the sound of that. She thought it would make a good t-shirt on people who had become weak and unable to work physically. Comfort had a price.

When Mari and Jean got home from the museum, they were talking a lot about Mari getting pregnant by the end of the year, and how Abdul would fly in to meet his new family. The new cyber couple would then drive back to the East Coast together. Everything was like a fairy tale come true. The Black Cinderella dream was the real deal! Back at home from the museum both went to sleep that night feeling unsettled about the divisionary docent at the museum. Mari went to sleep feeling bad about telling Jean that the indigenous woman was glad to get away from white people to be with color only instead. Jean resolved not to go back to the museum because what was the point of hanging out with the hate vibe? Her energy and passion was beginning to wane.

The next morning, Mari, with a heavy heart, wished to share with her white benefactor that she felt like the museum woman had been triangulating her way in between black and white. She assured Jean if it had gone further she was ready to stick up for the old woman. Jean felt reassured. The sacrifices that had been made promoting the young black widow's safety and wellness were appreciated. Or, was the black woman just using the white woman? It didn't really matter because everyone was stuck quarantined with Covid running the show. Jean could have thrown a tizzy and forced the black woman into homelessness, but what good would that do? Everything was riding along smoothly with the plan of Abdul arriving soon and Jean getting her life back.

It had been such a complicated and unexpected journey. She was looking forward to getting back to her white-privilege lifestyle that she had had before she stood up and did the right thing. John would declare she had tripped over her two left feet. Jean was determined to make it out alive, and Mari was in love with a philanthropic Iranian Turkish Christian from the other side of the world. Both women were giddy, riding on the high of talking about babies, fancy apartments in NYC, and Abdul wanting a Tesla. Things were good. Then the bottom fell out.

Abdul reported that his company was not cooperating with the plan, and the U.S. State Department issued travel warnings to Turkey for Covid and terrorism threats. Jean had based the next three months of her life on Mari leaving, and now God was fucking everybody again! Jean's blood daughter entourage planned on coming out for a two month visit because they were able to work remotely as young educated white-privilege professionals even though Jean had declared no visitors in 2020. Mom had already given the go ahead to show up for Thanksgiving and stay through the New Year, including a period of quarantine at the beginning. She needed Mari's space to be vacated, and Mari was sinking into another depression.

Their favorite place for talking things out had become the two padded bar stools in the kitchen. There was a long stretch of new butcher block countertop that ran along the west side of the kitchen making an L shape at the end which created a back rest for one, and a wall at the other end creating another back rest for the other. They were both comfy. Their seats were inter- changeable, neither had a favorite. They switched seats regularly depending on the circumstances, although in the beginning Mari was afraid to sit on either stool. She actually asked for permission.

Jean didn't believe in having a stove or oven because of the environmental load of mining for metals. There was a glass induction hot plate instead, and a little counter top oven that didn't heat up the kitchen during the summer. She had removed the A/C after her mother died because she hated the sound of it and the fuel it consumed. The carbon footprint associated with lavish and unsustainable living practices was not an attractive lifestyle for Jean. She preferred austerity with promoting tough and strong rather than excessive living promoting weakness.

The extra long countertop allowed both women a good stretch of space to talk out the dramas unfolding. Both had been born and raised in California, and both required a fair amount of personal distance, unlike other cultures where elbow room wasn't an option. Jean actually liked Coronavirus because the social distancing provided psychological distancing as well.

Being a person with compromised boundaries to begin with gave her a chance to step back from the world and not to be so available to every Tom, Dick and Harry who had a sob story to tell. Her white horse was getting a little gray in the muzzle, and the virus was an opportunity to put the old mare back out to pasture.

Bent on harvesting the silver lining of the black cloud called Covid, her experience with Mari, she prayed, would be her last hurrah as a rescuer. She prayed to God she was done helping others. In the past she'd gotten off on being the savior. She had finally worn herself out, she hoped. She was too old to rescue anybody else but herself now. Jean was accessing her own personal survival mode. She had places to go, things to do, and people to see, but mostly she had a planet to restore before she dropped dead. White privilege made the mess and it was up to white privilege to save the day. She was excited for Mari to be gone so she could get back to the business of being in charge of the Earth's wellness campaign. The people living in ghettos weren't in a position to restore the planet. They had not made the mess, but they certainly were victims of it. Jean secretly hoped that the inner-city dwellers would jump at the chance to get out of the cities to do the necessary labor needed to restore nature, like planting and weeding. Certainly the white collar HVAC types were useless to perform the real work!

With the turn of events of Abdul weighing heavily on both female hearts, for different reasons, Jean was now determined not to allow the energy shift to get a toe-hold into her foundation of strength. She was calling the situation, "Mari's Relocation Program". Abdul was Jean's ticket to freedom. She made sure that morale was kept hopeful, high, and moving forward. She went into her, " Okay, it's cool, not a problem, we can ride this bull in any direction mode." She approached Mari after the State Department's travel restriction announcement with the suggestion of, "Let's-Get-Organized!"

Jean had a strange habit of getting all jacked up on caffeine, and then becoming almost like Mary Poppins on steroids. Her fervor for getting everyone excited about the job at hand was contagious. She'd just start telling everybody how wonderful things were, and the next thing you knew they were all working away whistling Dixie. The problem was a false high created by Jean just to keep her own head above water. She refused to go down. She kept everyone buoyed up and moving in a positive direction, which was fine, but she probably shouldn't have gotten herself into the predicament in the first place. In her own mind, she had already given her big contribution to the Black Lives Matter movement. Now she just wanted to get back to the white privilege that the rest of her peers knew how to maximize because they had never known anything but white privilege. It would be like asking a forty ton whale to live in a bathtub. After all it wasn't the fault of the white folks that they didn't know how not to be spoiled with entitlement to burn. Right?

Jean, however, was remaining stuck with no respite on the horizon. Mari's passport was still being processed in South Carolina, which she tracked everyday, and Abdul was still in Turkey. No matter how much caffeine Jean drank she couldn't change the situation. Mari was still living at the house that Jean needed for her soon-to-arrive real family.

By this time, Mari had become amenable to getting on board with anything that Jean was requesting. She got all of her stuff tightly packed up and moved over to an empty shed at John's house, making her footprint in Jean's life greatly reduced. The goal was to get Mari down to one suitcase, one backpack, and one toothbrush. They both laughed when Jean suggested that Mari cut the handle off her toothbrush. Running to the bathroom and pulling out the latest environmentally approved model having a handle made of bamboo with a brush that could be changed out and composted made Jean love Mari more. Now Mari was the environmentalist and Jean was the student. Change on the horizon was looming large.

The Breonna Taylor case was unfolding daily in the news. The twenty-six- year-old EMT's family had already received a big settlement from the Louisville, Kentucky police, but charges against the officers had not yet been filed. Early every morning Jean tuned in to the CBS News on TV to hear the latest, and every day she would report to Mari the similarities between Breonna and Cedric. It had become routine for Jean to call the Suffolk County District Attorney's office on the twenty-ninth of every month to request a copy of the investigation. They were not returning her calls. Obviously, the case was still under investigation thirteen months after the alleged police shooting of her friend and Mari's husband. To Jean, the Boston city government was practicing avoidance.

But Jean was like a little terrier, she wouldn't let go. She couldn't let a sleeping dog lie. For months, she had been afraid of the race card that Mari held over her head just because she was black. It wasn't easy to make black friends if there weren't any to be found. Cedric and Mari were a first. But as the two women became more familiar with each other over time it became Jean's routine to discuss the difficult subjects that Mari preferred not to discuss-like the FBI and opening up a civil rights investigation. Eventually they completed Mari's statement of events from the night of the police shooting, and Jean wrote a proposed cover letter to the FBI that could be attached to Mari's statement. They talked about going on Dr. Phil or Oprah, but Mari would never even look at the bat, much less pick it up to take a swing at justice. Jean wasn't angry, but she was definitely frustrated with Mari because if the widow were to get the money she could move out and get her own place, which meant Miss Jean could get off the hook.

Utterly speechless that a wife whose husband was murdered by the police was not willing to collect millions of dollars and seek justice reform left Jean exhausted at the end of every day. It was this fact that finally made Jean get past her race issues and being consumed with white guilt. She didn't care anymore if she was walking on eggshells with her black friend. She told Mari regularly that she could feel Cedric from the grave pushing her, to push his wife, to seek a cash settlement for justice reform. They were at the point in their relationship whereby Jean was constantly telling Mari how bad white people were, and how they needed to get it through their insane white brains that people of color were human beings and not lesser animals!

They only had a couple more sentences to add to Mari's statement of events before it was ready to launch into play. Because of the delay problems with her passport, and the travel restrictions to Turkey, unemployed Mari, called up her old job at the hospital. They hired her back right away. Now she would have another excuse for not pursuing justice and a settlement. Abdul didn't want her to go back to work, he thought it wasn't safe.

Jean's friend Steve was blocking the driveway one afternoon when the two women arrived back home after being out for a day of fun adventure to the local ghost town. Jean parked the car on the street and got out to greet her friend. She encouraged Mari to get out of the car to meet Steve also, but she didn't. She just sat in the car waiting patiently. When Jean finally realized that Mari wasn't budging, she opened up the back hatch and said, "What's going on in here? Come on out. I want to introduce you to Steve." Mari gave the excuse that she was worried about her hair not looking good, and Jean just shined it on. Later, when Steve and Jean were talking on the phone he admitted that he thought the thirty-six-year-old Mari had acted like a teenager because she was unable to conduct reasonable social graces.

It wasn't a problem for anyone. Mari ended up saying hello and Steve also gave a well-mannered greeting, but he mentioned it again when he and Jean were chatting on the phone that big weekend before the presidential re-election of Donald Trump. Jean found it so curious that Steve was being somewhat critical. It felt to her like Steve was saying that Mari was an immature thirty-six-year-old. Since the two women had developed a close mother-daughter bond, Jean was willing to try to rationalize Mari's behavior to Steve. But after she got home and started thinking about the cultural differences between those living white privilege and those not, she was contemplating the possibility that Mari, being a woman first, and a person of color second, was perhaps just being polite and reserved instead of assuming that the thing to do was to jump out of the car and throw out a hand shake. Covid or not!

Mari's semi-different behavior, for Jean anyway, was a bit of a game changer. Even if Mari were awarded six million dollars through the court system she would still not want to live on her own. White people in the USA had one kind of life, and people of color were different. White women knew how to shake hands with anyone. Jean remembered back in the old days when she was about the same age as Mari, she didn't know how to shake hands either.

She had to force herself out of her comfort zone and, basically, man up. She remembered back to the eighties when men didn't know how to shake a woman's hand. If it did occur it was awkward. Mari simply didn't know how to do it the modern white way. Jean had surmised.

Jean was asking herself, why does a certain way have to be the wrong way just because it's a different way? Why do white people think that their way is the right way and that Mari's way is immature? John uttered the same sentiments as Steve when Mari had jumped ship from the white cop and took up with Abdul in Turkey online. Jean realized that educated white woman in America had a different ride than uneducated women of color. It was profound! The white woman could write her own ticket and not bat an eyelash. Whereas, the women of color were still functioning on the old system of depending on a man. Or in the case of Mari, she depended on her sugar mama, but was actively on the hunt for a replacement. Jean was being used and she knew it. She didn't mind because she was making a substantial contribution to Black Lives Matter, and still thought she should get a line- item tax credit for service to society.

In the beginning when they had first started out, Jean found it difficult to live in her own house. She'd spent a great deal of money to have things the way she wanted them. It wasn't fancy just reasonable and decent, and she wanted to keep it that way. She tried to give Mari a chance to pitch in and help out but every time she did she felt this uncomfortable vibe coming from Mari. She felt her unwillingness to pitch in and help. Because Jean had deep psychological issues stemming from not being loved as a child, she just stopped asking for help. She decided that offering Mari a free ride was a lot less stress.

The old white woman felt no anger or animosity, she just chalked it up to Mari being black with issues around helping white people. Jean felt like she was living in God's laboratory, and that she and Mari were God's lab rats. God was giving these two women a chance to change the world. Jean was inspired to comply, so she lived in her house on egg shells while waiting for Mari to be on her merry way. After Mari dumped the cop and was planning on moving to Turkey Jean finally pulled up her big girl panties and told herself, Enough! I am getting out of this alive and back to my good old fashioned white privilege. She rationalized, there was a reason why white men no longer wanted white women. The educated white women were no longer willing to be subordinate to men because it was no longer necessary for them to do so. Mari knew how to play a man up one side and down the other, and so did Jean, but she didn't have to. Jean was self sufficient. Mari was not.

Everyone settled themselves back into the lull of waiting. There was nothing more that could be done to get Mari gone and on her way. Jean could've kicked her out, but that wasn't going to happen because she loved her. The two women had bonded through thick and thin. Jean was sitting up in her bed typing on her laptop, Mari had just left for work at five AM. It was election day 2020. The news had shown the night before that places like Rodeo Drive in Beverly Hills and Park Avenue in New York were being boarded up with plywood because of the expected violence. That's when the call came through from the Suffolk County District Attorney's office. They were calling Jean because she had been the one placing all the calls.

The person calling was a woman. She wanted Jean to give out Mari's phone number. Jean refused to do that because it wasn't the right thing to do, but she was willing to contact Mari by text to put the two women in touch with each other. It was obvious the woman was trying to pump Jean for information, but she was nice about it. She wasn't brutish or anything like that, and Jean was more than cooperative. Because Mari and Jean had talked so much about how to proceed, their ship had finally pulled into port. Mari's unwillingness to proceed was finally paying off. Had it been left up to Jean, she would have called out an air strike on Boston and brought in her own dogs to do battle head on. But because of Mari's inclination to listen to God, or maybe she was just scared to death of the government (the police), finally thirteen months after Cedric's death, it appeared something was finally going to happen!

The woman from the DA's office and Jean chatted for about fifteen minutes. Jean probably said too much, which was typical. She related, to the woman, information about Cedric's PTSD. She told her about the marijuana smell that may have triggered his attack. She told how she had become friends with the young black couple, and that she completely understood how it was impossible for the DA woman to divulge any information. It was all up to Mari. The only thing Jean had done, for Mari, was not to let the ball drop. Jean told the woman that she and Mari had taken three months to write the statement of her experience of the killing of her husband. Jean did the best she could to tell the woman that Mari was not after revenge. Because of Jean's inability to let it go she was becoming the white "fight" to help her black friend seek justice. That's what she told the woman.

Again Jean was involved in something that was none of her business. But if Americans didn't stand up for fellow Americans, what good could we possibly be as a nation? She had some satisfaction. All of the inconvenience was worth it if Mari could be reasonably compensated with a few million dollars. The time had come for Jean to start wondering about how the cash settlement from the Boston coffers would end up transpiring, and how much it would be? At this point, Jean and Mari didn't cry as much as they did at first. Mari just wanted to get on with her life. And so did Jean.

Even if it were to turn out that Mari was unable to make her car payment and pay Jean back, there was a profound sense of immeasurable satisfaction for the older white woman. It was a connectedness to a bigger purpose. Jean enjoyed having a ringside seat in the changes that were occurring in the world, and especially in the country that Cedric had fought to protect and defend. This was her true privilege. Jean wasn't going to take her old white mare back out of the pasture and ride anywhere else ever again, she hoped, but she sure was happy that she had that horse to ride this one last time.

Jean told the woman from the DA's office that she had been listening to the news of Breonna Taylor. The representative of the law drove home the point of how important it was for Mari to contact her before contacting any legal counsel or the press. It made perfect sense to Jean. Good old God was going to come through again. Jean's nose would start to itch if she was getting teary eyed, and her nose was itching up a storm at the thought of seeking and receiving justice for Cedric. Jean had called the DA's office on the twenty- ninth of October just like clock work, and four days later they were getting a call back. It had been thirteen months. Jean was feeling like a hero. She wanted to paint her white horse black just to celebrate.

The strange part happened a week or so before, on the CBS morning news, Gayle King told of the police being accused of "Wanton Endangerment" in the Breonna Taylor case. That was a term Jean had never heard before. She looked it up on Google. She was anxious because every time she got a new piece to the puzzle she would report it immediately to Mari. Her internet search brought her straight to an attorney's website which had the information she was looking for, as if someone were just sitting there waiting for the next victimized family to call. The definition of Wanton Endangerment fit Cedric's death. It was the same as Breonna Taylor's.

There were B&B customers staying in Cedric and Mari's apartment, with them, at the time. Had they left? Were they still there when the shooting took place? What about the dogs? It seemed to Jean that the concept of Wanton Endangerment applied to the circumstances surrounding Cedric's death as sure as could be. Both he and Breonna had died from gunshot wounds fired by over-zealous police.

# CHAPTER TEN

# The Rules

Afew days had passed when she asked the young black widow if she had finished the review of her eye witness statement. At that point the two women were still planning on, or at least discussing, contacting an attorney who was looking to become famous. Since there seemed to be a run on social justice for black folks, Jean felt that Mari should join the crowd, and get in line for a settlement just like all the rest. Change was not going to occur overnight, but it would never occur unless a demand for justice was sought by those victimized by abuse-of-power. Mari stated that she thought her statement was good-to-go. Occasionally, the two women would make time to go out for a meal and take in a little adventure, but since Mari would soon be leaving for Istanbul they realized that their time together would soon be coming to an end. Even a trip down to the dollar store was an adventure for those two. There was a spiritual bond that words could not describe. Jean felt they were finally on the same page, and that page was justice for all. Jean liked treating Mari to an outing because she was trying to encourage her to do the right thing: Seek Justice!

Carmine B. Littleworth

At The Home Depot, Jean needed to buy a couple of gallons of interior wall paint. They didn't have time to go out for lunch. They were just out running a couple quick errands because Mari had to get ready to go to work. Jean wanted to tell her something about that conversation she'd had with her friend Steve regarding Mari's immaturity. It wasn't easy to tell your pseudo adopted daughter about someone else's opinion declaring them sub-standard. Jean thought she could pull it off without any hurtfulness because she wanted to discuss how humans, specifically white males, decide what is proper and what is not. Jean was perplexed with how white people still made all the rules.

She related that her friend Steve had decided that Mari, by not jumping out of the car to greet him, was immature. Why does he decide what is the right thing and what isn't? The year 2020 will live on in infamy for centuries to come. It was a year of Covid, Black Lives Matter (BLM), fires, divided politics, earthquakes, hurricanes, and not to mention the anniversary of women's right to vote being shoved under the carpet. Jean couldn't help but wonder when the locusts would show up to eat all the crops. Goodie for the rule makers. Yippee. Wrong! Jean pursed her lips and blew out in semi-desperate frustration with humanity's lack of vision for the future.

She had always been a huge fan of white males. Her beloved father was a white male. Her brother was white, and so were her three white ex-husbands. She resisted the temptation to bash white males or any male for that matter. But it did dawn on her in this one instance: White males make all of the rules and the rest of us can only kowtow. She reflected back on the book she had read in high school, Soul On Ice by Eldridge Cleaver. White Males vs The Rest was the story line of this required reading for her California high school humanities class back in 1972.

Because of Jean's frustration with white people abusing their power, not being kind, and taking the lion's portion of the natural resources it dawned on her to find out the percentage of white people on Earth. She Googled it. There was no readily available number to be found. She did her own research to find the world percentages of ethnicities. She added up the population numbers of the colors of the world with her pocket calculator. Her findings revealed only seven percent of the world was white, and the rest were people of color. That said to Jean that just three and a half percent of the world was running the show and making the rules just because they had white penises. Seemingly, that would not be a problem for Jean if the rule makers were trustworthy, and did not practice abuse of power. But, in defense of white males, she also felt that white women, in their constant lust for things, had put the white males up to it. Maybe it was a scientific phenomena. Maybe the white males in their natural-lust to get-laid were victims of the innate desire to procreate. Jean didn't want to throw stones in anybody's glass house, she just wanted the planet fixed. ASAP!

She really had a hard time admitting all of this to herself because she felt like she was being a traitor to her white male father. Her dad had been dead for sixteen years. They were best friends her entire life until he passed away. When her dad entered the old age and frail stage Jean was right there to be his constant caregiver. She had been out hiking with another one of her white man pals in an area where she and her dad had hiked for many years back in the day. It brought back so many fond memories, but she realized she had not been thinking much about her dad in the last few months because of the up and coming election between Donald Trump and Joe Biden.

Dad would have supported Donald Trump. Jean appreciated Trump because so many positive changes had occurred for people of color under the Trump administration, such as the Black Lives Matter movement. She felt all of the positive inroads that needed to occur happened under a conservative white male rather than a liberal. Changes under a liberal would have brought civil war, but changes under the extreme white wing came as inevitable and not from the liberal bent. She thought Trump was a classic case of concocted, trumped up, reverse psychology. Trump was bringing the Proud Boys out of the woodwork and exposing them to the light as their leader. It had to be a pre-planned event. There was no doubt in Jean's mind, but she was the only one who could see it. Donald Trump was a necessary agent of change.

John and Jean tried talking politics. John would scream at her for appreciating Trump on any level. She saw that Trump had done a magnificent job of exposing American white supremacy to the entire world. John saw him as a monster. Those militant Proud-Boy faces were been caught on camera. People were now identifiable. Has computerized face recognition become the savior of humanity? She asked herself. Maybe if there were enough decent Americans demanding justice human evolution might stand a chance? Jean pondered on this long and hard.

The week before the election while Mari was back to work at the local hospital she felt the extreme pressures of being high risk. It was not an easy job with or without the pandemic, but Coronavirus brought on a whole new ballgame. She was the intake person at the front desk who took everyone's temperatures and interfaced with John Q. Public. She even had to interface with the people who were unwilling to wear a mask. Calling security was her job. Mari was one of the brave Americans leading the charge for public safety. This in and of itself was an issue for John and Jean, older Americans in the higher risk category. John was especially high risk because of asthma and a heart condition.

When Abdul and Mari were planning on being able to meet for the first time and leave the compound before election day, Mari quit her job because Abdul didn't want her working in an unsafe environment. Everyone was much happier because she was no longer taking unnecessary risks. But then the bottom fell out with the terrorism in Turkey, reduced diplomatic relations with the USA, and Mari's passport being stuck in South Carolina, so she decided to see if she could get her old job back. She was very polite and had asked for Jean's permission to go back to work. If Jean wanted Mari to make her car loan payments she had to say yes, but was sad that it had to be. So close yet so far. Jean's plan had failed. Mari was still in residence and now back at work. Of course Jean wasn't going to tell Mari not to go to work. There was not only a car payment to make but also just sitting around and getting bored could lead to other problems like depression. Jean told Mari, "Do what you gotta do." Mari did. She went back to work at the hospital and they were glad to have her back.

Early on when she first started working Jean was so proud that Mari had set her goal and accomplished what she set out to do. She worked temporary jobs just to put a few dollars in her pocket, but she was able to hold out for the higher paying hospital job with full benefits, and so close to home. Jean's blood daughter was also employed. She was able to work remotely from home with a low risk to Covid. She had a college education. Mari did not. When Mari came home that week before the election to describe the circus going on in the hospital parking lot Jean listened with a protective ear. The younger woman related that there were beds of pick-up trucks filled with white males, wearing guns, donning Trump flags and circling the parking lot in front of the hospital. Mari felt afraid. Jean was afraid. Abdul was afraid. The intimidation was successful.

Jean watched the news and saw that this was happening all across America. She thought of her dad, who would have definitely voted for Trump, and felt very conflicted. Jean was happy that Trump had exposed the white supremacy movement. It looked like it was inadvertent, but she didn't think so. She saw it as all trumped up. She felt it was a well orchestrated plan by the government, both sides of the aisle, to change American society for the better. America had dropped in the world, but was still a bastion of hope. Trump was an important factor in getting white America to change its ways. Trump was the government's attempt at reverse psychology to get white Americans to adjust to a multi cultural world. Trump was a necessary evil for America. Steve, the one who saw Mari as immature, thought Jean's politics were from outer space and regularly told her so. She was out there all on her own.

She saw how Trump had been groomed for service over the past thirty years in an effort to cooperate with the world putting the United States back on the map as an up-dated world player. Being a nation of freedom and justice for all with technology and green industry was clearly an understandable vision. The news media had pegged Trump as an evil white man. The news media had done a great job of encouraging a massive divide between left and right. People of color were now being supported publicly by white people. In one town during the Covid summer they had a Black Lives Matter demonstration, but no black people showed up. There were only white people demonstrating and protesting on behalf of blacks but against them as well. There were Trump supporters tussling with white BLM supporters in the streets. White people were fighting white people. The black people were too smart to show up for that kind of an event.

The day after the election was the day the woman from the Suffolk County District Attorney's office called. Jean could have approached Mari the night before to ask if she had returned the phone call from the DA's office, but the evenings were not her strong time and instead she cuddled up in bed to watch the election results. Mari and Abdul were having a marathon phone conversation in the next room. Jean had the TV turned up. She knew better not to intrude on the budding romance since Abdul was Jean's ticket to freedom. The next morning she jumped out of bed, downed a whole pot of tea, and approached the young black widow in the kitchen. When Jean spoke Mari's name, she responded in a way that let Jean know that her time was short and the subject was delicate. Mari only had time to get into her car and get off to work. Jean knew to be brief.

"Did you call that woman?" "I did." "Did you leave a message?" "I did." "Did you let her know your schedule?" "I couldn't because my schedule is so mixed up." With those words Jean knew that Mari's feet were still in drag mode. Jean assured Mari that whatever she decided she would back her up. There was no other option. Mari had to decide for herself. If Jean were after money she would have been pushing for fast justice, but she wasn't. She just wanted to get out of the predicament alive, and get back to some good old-fashioned white privilege, where she could live out the rest of her life without all of the black drama.

Steve and Jean went out for lunch the Wednesday after the election. She told him she had received the call at six o'clock in the morning west coast time. The DA's assistant had tried to pump her for information about Mari. Jean explained how it appeared that Mari was stuck in neutral and couldn't move forward. Jean was no longer frustrated with Mari but rather challenged by the task of getting her to kick the ball into play and get the show on the road. She wanted her house back and she wanted the young widow to have a life outside of turmoil. Steve, who had been following the drama from the beginning as part of Jean's inner circle, came right out and told her exactly what she already knew.

When she got back home that afternoon she fed her dog, checked the mail for Mari's passport, and then settled down to read email. She was excited to see Mari's face peek around the corner of the northwest side of the house, where Jean was basking in the sun on some of the last warm days of the year. The sound of gunfire perforated the calm. Shots rang out. Both women heard it off in the distance. Jean had heard it for the last forty years. It was duck season down at the lake. She paid no attention. Conversely Mari was terrified. When Mari peeked around the corner Jean was excited to see her dear friend who, hopefully, had the information that was needed to push to the next step of getting the show on the road with Cedric's Justice.

Jean smiled and motioned for Mari to come and join her in the delicious warm sunshine. She could see there was panic on her face. She was wearing her warm and comfy house clothes with her hooded sweatshirt pulled up. Mari had lost a great deal of weight since moving in with Jean. She was looking pale, almost ghostlike or ghoulish. Jean was no longer willing to be distant from Mari psychologically. Things had transitioned into a full blown mother-daughter honesty. A certain safety settles in when a parent is always on guard with a child. It is unwavering. It is constant. It never questions motive. These two were close to that level because Jean felt needed. Both had their own trepidations, but both were intertwined on a spiritual level. If God were to take one of them at that very second neither would have regretted being shoulder to shoulder in life . This was true especially for Jean, who was so much older, and Mari was grateful for the shelter.

"Get over here, right now!" Jean spoke to her scared child. Mari had tears running down her beautiful copper-brown fully rounded cheeks. "What? What is it?" Jean thought she must be upset about the phone call from the DA's office. "I heard gunshots. I was asleep. Is everything okay? Are you okay?" Mari questioned in an anxious manner. "Yeah, yeah. No problem." Mama Jeannie explained about duck season down at the lake. "I was so afraid! I thought maybe someone was on the property killing the deer, or maybe your dog got shot." It was really obvious Mari had PTSD from the night of her husband's killing.

There was just a half hour of warmth left on the clock before the sun set behind the mountains. Jean wanted to tell Mari what

Steve had said. "I told Steve about the phone call. I told him that you were having a hard time moving forward, and it's like you are scared to death to do anything what-so- ever." Mari kept wiping the tears away. Jean was on a roll. She was going in for the slam dunk. "Steve said I need to just take over and run the show if you are unable."

Back on that sunny day before the election, when the women were shopping at Home Depot, the shift that had occurred in their relationship while they were standing in the paint department was a game changer. Jean asked Mari if she would go get a shopping cart. Prior to that she had been too afraid to ask Mari to do anything because of what she perceived was a black person unwilling to help a white person unless forced to do so. This was a huge breakthrough in their friendship. They were feeling safe with each other. The paint was too heavy to carry and Jean had forgotten to grab a cart on the way into the store. Mari was on it immediately. There was no hesitation. She came right back and Jean loaded a gallon and a quart of mis-matched paint off the oops rack that was cheap, but in colors that would work easily in the house. She also needed the expensive paint mixed for the front porch trellis that John was re-painting on the house next door. Jean said, "Let's take a walk." It would be five minutes of waiting for the expensive paint to mix. Mari pushed the cart. They cruised the aisles so that they could continue their conversation about Steve thinking Mari was immature. Jean wanted to make sure Mari understood about white people, white males especially, who make the rules and decide what is proper behavior for everyone and what is not. Steve had decided Mari was immature by not jumping out of the car to greet him. Mari didn't want to butt-in. That was her defense.

But there was nothing to defend. It was a system that had been in place for thousands of years. White people make the rules. It was so ubiquitous on Earth no one noticed, they just accepted. Until, one day, the BIPOCs all woke up collectively, during the pandemic, and complained about the white abuse of power, especially in the United States of America. Being black in a white world is a navigation much like that of the early explorers. What's next on the horizon may be easy, or it may mean death. When the two women talked in the sun at the back of the house during duck season, Mari admitted she was afraid of the people who had killed her husband. She felt like they had a motive to want her dead. She was scared to death to move forward. It had nothing to do with being immature. It had everything to do with not feeling safe in the white world.

It was valid. Jean couldn't argue with the young woman's logic. The District Attorney's office was missing one statement. Mari's. What had America become? Too fearful to seek justice? Jean's father had taught her to be a woman of principle. When she was feeling strange about not keeping her dad close in her heart because of the election, she wondered if her dad would've supported BLM. How could anyone not support BLM? The principle was clear. Thou shalt not kill. How difficult is that? If the Ten Commandments need to be re-written then let's do it. But if the Ten Commandments are still valid then why not follow them? Simple as that. Jean had no idea how the world would turn. All she knew was that she had a principle: Liberty and Justice for All.

Two days after the election the results were still inconclusive. Donald Trump was suing a few states for illegal ballot counting. There were a lot more pick-up trucks driving around with American flags and Trump banners, and gun brandishers not wearing seat belts as they stood up waving their guns in the pick-up beds behind the cabs. Jean was telling her real daughter on the phone about the trucks. "You mean they look like Jihadists?" They roared in laughter. Her daughter was a chip off the old block when it came to setting a visual for the mind. It put things in perspective for the mom with two daughters.

Jean regularly followed the social media platform Nextdoor to keep track of their neighborhood. On a thread from a neighbor, another woman was also feeling intimidated by these types of truck drivers. A white male replied that it was perfectly legal to flash your guns, wave your flags, and state that Black Lives Matter was evil. America was divided. White people were feeling threatened by BLM! What exactly is the threat that the white people feel? Jean thought about America's principle of Liberty and Justice for All. Was it just a white wash? A scam? Why couldn't America uphold the founding principles? Was it that the white people feared they were going to have to pay for reparations, and they didn't want to have to cough up the bucks?

There were a lot of thoughts running through Jean's head simultaneously. How was she going to get Mari safely back on her own two feet in a world that wanted her dead so they didn't have to pay out? That was her big concern. She loved the young woman, but she wanted her life back. How was America going to bring up the standard of living for all of the people living in poverty? Who was going to pay for that? The second question was a bigger concern. She knew that in many ways Mari was like a child that needed protecting and that for some strange reason God had decided that Jean was the soft-hearted woman who could at least try to get the job done. Mari wasn't a huge problem to live with, although it was easy to complain about little things like the smell of all of her beauty products and laundry soap, which were toxic for both Jean and John.

Since the bigger picture was restoring planet Earth, and preventing the extinction of the human race and other living creatures ignoring the self, in order to promote the larger theater, came naturally for some. Jean felt if everyone could love one another then perhaps restoration of the planet and evolution was a possibility. The fighting and the killing of everything had to stop.

Carmine B. Littleworth

The human race was being asked to get its shit together or die. What was so difficult about such a basic concept? Jean wondered this until she was blue in the fucking face.

# CHAPTER ELEVEN

# The Neighbor Across The Street

To the chagrin of some, Biden was declared president-elect on the first Saturday after the election. Jean congratulated all of her Biden friends. She herself had voted for Trump, even though she wanted Biden and Kamala Harris to win the election. Whoever she voted for never won, so that is why she voted for Trump. She never thought about her vote until election day. She just knew for some reason she had to vote for Donald Trump. After Biden won, she knew why. She was planning on the possibility of taking on the entire white male American racist population if it were necessary. If they were to discount her as a wacko liberal, she could shrug her shoulders and say, "Hey, I voted for Trump." That put her in the position of being able to sleep with the enemy, which she had done regularly throughout her adult life.

She congratulated Mari on her vote, through text, and they continued with their conversation after Jean returned from her hike. Mari was cooking breakfast. Jean made tea. In the neighborhood where they lived there seemed to be an element of gun-toting-negro- haters. The best bet was to get Mari gone. Mari talked to Abdul about it and he said he didn't want his girlfriend to speak to the District Attorney without him being present because he needed to be there in order to protect his woman. Jean saw Moses parting the sea again, and fantasized about running down to snatch up a fish to eat all for herself. Her plan was working. Abdul was taking over and Jean was closer to being off the hook. These were trying times for all humanity.

The two women talked about how Mari was going to be the star witness in the case and Jean would be the character witness. The dining room table deposition, which had been taken over a three month period, would just have to stay on Jean's laptop. They agreed: Let a sleeping dog lie. They were both afraid of violence from Boston and violence from local white supremacy. Jean had become a white person afraid of white people.

Gears do shift in a healthy transmission. Now all they had to do was wait for Mari's passport to arrive. Abdul was in charge now. Someone had flipped the switch to on in the tunnel of darkness. No longer would Jean be shooting her mouth off on Nextdoor. A quiet fear had crept over the valley where they lived. It seemed like people were wondering if the Trump supporters were gearing up to go on a killing spree. Jean didn't want to find out. Secretly, she just wanted to mosey on back to the quiet life she'd had before Black Lives Matter and the Proud Boys had perforated her quiet privileged bliss.

It was slightly disconcerting for Jean that after everything she and Mari had been through together Mari saw Abdul as her savior. Abdul was going to ride in on his white horse to save the day. It reminded Jean of a conversation she'd had with her real daughter about narcissism and the Proud Boys just after the election. She had drawn a parallel for her daughter, who was savvy about the world. She asked her daughter, "Who is more mentally ill? The narcissist or the narcissist enabler?" Jean didn't give her a chance to answer the question. The enabler was the answer. "The narcissist is fed by the enabler. The enabler does all the work. We, enablers, have set up the narcissists to run amuck. We, Americans, have made fools of ourselves in the world, hence the birth of the Ugly American. We make up ten percent of the world population and consume ninety percent of the resources. These figures are probably not exact, but rather used to paint a picture with a broad brush." Jean was tired of assholes running the world! "The planet is dying and people are living in unstable conditions. The poor people aren't the problem. It's the rich! Their lifestyles do not represent sustainable living!" A lot of people didn't want to hear what Jean had to say.

Before the pandemic, when Jean and John were attending a Christian church, the congregation had been asked to help out with the homeless population so that people in the streets would stop dying during the cold of the winter months. John was chatting with a conservative couple after the service. "Let 'em die." That was the sentiment of the good Christian church-goers. The problem with that method is disease. Diseases are living organisms that are non-selective when it comes to socio-economics. If science were capable of culturing a disease in a test tube that could take out just the inner city scum that plagued the world then wouldn't that solve all the problems? The Proud Boys wouldn't have to pay for reparations and the inner cities could be bulldozed. Then new construction would get real Americans back to work. Was this how people really thought? Jean cringed. Maybe that's the way nature made white people. Maybe white people have to be more aggressive to survive because white people are the physically weaker color? I wonder if that's true. Jean asked herself.

White males make up approximately three and a half percent of the world population and are desperately trying to maintain their power. It's a lot like holding onto sand. It just keeps drizzling out the bottom. Jean was done with the American person not willing to share and fix the mess that her color had made for itself. She joked with her friends, " I want that tax credit for running my own BLM shelter facility." They laughed. She wasn't serious, but she was. How will America fix itself and who's going to pay? She refused to lighten up.

John and Jean stopped talking politics because Jean saw Donald Trump as an agent of change. He was a human being on planet Earth who was willing to go out there and make a fool of himself because America was in deep yogurt across the world stage. Tourists from Europe were too afraid to come to America for a vacation because they were afraid of getting shot. America was a nation infamous for its gun toting rednecks and damn proud of it! Jean was tired of the narcissists running the show. She was even more tired of the role she had played in enabling white power to create this unsustainable world.

Years back when she was on the social media platform Linked In with the scientists from around the globe, the consensus was, "We gotta back outta this one real slow." Jean added, "It would take every man, woman and child on this planet to turn this bus around or face extinction." If white people made the mess, and the people of color are being asked to clean up the mess then wouldn't it be a good idea for the Proud Boys to go sit in the corner for a few years and either eat a slice of humble pie or shut the fuck up?

How much more bullshit coming from the USA was the rest of the world going to put up with? Jean was always trying to explain that Trump was the vehicle for getting white Americans to see the unstableness of the current system. It was also the platform for America to show the United Nations, "Look at us. See how good we are behaving now. We had just elected a woman of color as Vice President." Often times, Jean would say to John, "We have Donald Trump so that we don't have to have the UN tanks driving down our streets declaring martial law." He would always reply, "That will never happen in America." But why not, if Americans can't behave? You could easily say, much to the chagrin of some, that the Proud Boys were on the verge of destroying America. That old Jean Stinson, she was a thinker.

She couldn't help but wonder if her dad would have secretly liked the Proud Boys since he had come from old American stock aka the Mayflower crowd. Dad's people were not slavers, but they were Indian and buffalo- killers. The argument that white folks liked to use regarding Black Lives Matter was that their ancestors didn't own slaves. John, a relatively new immigrant, liked to use that one. His Dutch people sold rubber out of South America. That didn't hold much water with Jean. She was one of those who didn't separate out all the parts. She always saw everything as a whole. Until the Proud Boys could escape to another planet we were all stuck in the shit show together.

Mari and Jean agreed to let the sleeping dog lie and not proceed with the District Attorney. Abdul decided he needed to protect his woman. Jean just chilled. She went into kick back mode; skate and wait. She was patiently biding her time waiting for Mari's new passport to show up and then freedom would be hers. She could taste it, smell it, feel it. It was coming!

One Friday Jean noticed that the neighbor across the street, the one who would wear his gun belt and six shooter down to the mailbox, had not brought in his trash can after the pick up, which had occurred on election day. She walked over to see if she could find him. The windows were wide open with the first snow storm of the season coming in fast. She yelled his name in the window. No response. She knocked on the door, no response. His TV was on. His light in the hallway was on. She hoped that was a good sign.

She walked back down the driveway, and went over to John's house to report. She asked him his opinion of what to do, "Should we call the police?" "No way, don't call the police on him. He's probably just passed out. He'll come out of it." John had battled alcoholism in years prior, so he knew what he was talking about. Jean trusted him on his judgement call. She knew the neighbor had a drinking problem. She had followed him out of the neighborhood one day, three or four months earlier, swerving like a snake. He was a menace, but she didn't want to hurt him by turning him in to the sheriff. She probably could've saved his life if she hadn't been so afraid of turning him over to a system that criminalizes health disorders.

Instead, she waited all day Friday to see if he would close his windows because of the cold. Throughout the night she would check to see if his lights were still on. They could be seen from her house across the street kitty corner. It was very cold and the wind was fierce that Friday after trash-pick-up- Tuesday/Election Day. She wondered if stuff was blowing around in his house.

Saturday was miserably cold and the windows were still open and the light was still on. Jean had a key to the man's house. He had given it to her years before in case of an emergency. She didn't want to go in and use the key because she was afraid that he'd be sitting there drunk out of his mind and shoot her for trespassing.

Finally, on Sunday morning, she couldn't stand it any longer. They had received six inches of snow over night. She'd just finished shoveling her driveway and the time had come. It was time to call the cops. They arrived. She met them in the street. She told them she had the key. She handed it over. The key didn't fit the front lock. The police had to go through the doggie door and in through the garage. They came back down the driveway a few minutes later and handed her back the key. Sure enough he was dead.

Jean didn't want to hold on to the key. She was afraid of the responsibility because she knew he had beaucoup guns. The deputy reported there were loaded guns throughout the house, and a safe full of guns in the garage. Jean didn't want to know how he had died. Her brother had died years before. He was all alone up at the family cabin in the mountains and dead for days before their parents found him with the help of law enforcement. She was clearly feeling burdened by the death of her neighbor.

John and Jean were talking in the car on their way up to the cabin to check on the heating system the day after the coroner removed the body. Even though they couldn't discuss politics together it always seemed to circle back around. Jean said to John, while driving on the ice cautiously as he was criticizing her for going too slowly, "Somebody needs to do a major study on the psychology of fear, and present it to the American people." It totally went over his head as did most things that came from her. He was certainly not a Proud Boy. He was a self declared Christian Social Democrat. She wondered if her neighbor would have been a Proud Boy had he been healthy enough. He had the guns, he flew his flag, and he drove a pick-up truck, but it was an old truck like John's.

It was no wonder they called it the Bureau of Alcohol, Tobacco and Firearms. Jean was lost in her own thoughts. After she snapped back at John for criticizing her driving they rode in comfortable silence while the speeding cars, who didn't know the shady spots of the road as well as she did, were applying their brakes up ahead. "Better safe than sorry," she muttered to herself. It gave her time to wonder why some people take fear to the extreme level. Why are the Proud Boys afraid of the Black and Indigenous People of Color? She was terribly deep in thought on the subject. John, lost in his own thoughts, suddenly blurted out, "The Proud Boys are cowards!" "How could that be?" She blurted back. " What a great question! Are the Proud Boys cowards?" Hmmm. They both mulled it over in silence.

Would the Proud Boys balk at the suggestion that they were afraid or would they agree? "What is the deal with the Proud Boys driving around the neighborhood with their flags, guns, and pick-up trucks?" Jean and John were back on dry sunny pavement now. They were talking again. Jean said, " It's like pick-up trucks have become the symbol of evil." She knew she was pushing his buttons a little bit because he drove a pick-up truck that he had purchased new almost thirty years ago, but she rescued the conversation before he had a chance to become defensive, "It's not all pick-up trucks. It's not like your truck. It's the ones with the tires hanging out past the wheel wells, and a set of steel balls hanging off the rear-differential. That's the quintessential Proud Boy chariot!" John agreed. "Pirate Trucks! That's what they are, Pirate Trucks! Those guys are like modern day pirates." He continued to describe the color of the pirate trucks as black or charcoal grey but always dark in color. Jean was just kidding about the steel balls. That had been a fad a few years back. It had either needed a law to make it illegal, or maybe the boys got a dose of common sense.

Jean wondered why the Proud Boys weren't driving around in white trucks instead so they could look more like the KKK, but that wasn't any of her concern. John was clearly bothered by them. He had been bayonetted in the butt back at Kent State in 1970, when students had been shot and killed by the National Guard. He was more on the hippie side of the spectrum. He was definitely not a Proud Boy wanna be. Jean secretly wondered if John was afraid of the Proud Boys. She knew she was.

The neighborhood communication chain was alive and well. Jean got ahold of her friend who knew the dead neighbor's ex-wife, who knew the dead man's sister back in Florida. The next thing Jean knew she was going out to the virtual barn to grab a halter and walk down to the virtual pasture to take the old white mare out for another ride. There was a big difference between helping your neighbor with a short trot down the street and a full blown ride across the country with a U-Haul trailer, and sharing the barn, her newly remodeled house, with others. Jean rationalized her involvement. She was happy to help the family of her dead neighbor. Unfortunately, she'd given the key back to the cop, not wanting to be responsible for the guns. Since she had turned the house key over to the police she had to get the key back from the county examiner to check the neighbor's house for heat so the pipes wouldn't freeze.

Winter was upon them. They were having record low temperatures the week after the death. Jean had no idea if there was heat on in the dead man's house, and no way to check without the key. The sister in Florida and Jean became connected through texting, and it was time to put this project on the front burner. Jean was a natural at helping; she loved it. Too bad she was too old to join the Peace Corps and go to Africa, but she was too old, and she didn't want to go anyway. With the fear of disease instead, she wanted to know what it would feel like to retire from her selfless giving. Sometimes she fantasized about becoming the wife of a Proud Boy. She might even be willing to start coloring her hair again and wear a push-up bra just to drive around in a pirate truck. She scoffed at herself. Don't be ridiculous!

Back in reality, she found her big concern was to get into the dead neighbor's house to make sure the heat was still on. Jean explained all of this to the poor old broken-hearted sister in Florida, who couldn't possibly understand why her brother had ended up out West in the cold country that he had dearly loved before he died. He told his other close neighbor that there was no way he would ever go back to the East Coast. Through all of the texting back and forth figuring out the heat and the key Jean sent a series of photos. They depicted the condition of the inside of the house so that the man's relatives would know what to expect when they arrived to deal with the situation. The sister sent back a series of old family photos. The first one was the man looking all spiffy back in the day. The second one was the sister and brother taken somewhat recently. The next was the one with her neighbor and his elderly dog at the healing service at church. The last one was the neighbor with his arm around a woman of color. A negro!

What? The neighbor was friends with a BIPOC or maybe even related to a BIPOC? "Wow!" Jean was so happy she was wrong about her neighbor. Maybe he wasn't a Proud Boy after all. He was actually a wonderful person? Jean wondered. The dead neighbor's sister had texted to Jean a photo of her wife and her brother together. Perhaps her neighbor had accepted and loved his BIPOC sister-in-law. The sister told Jean they had been married for thirty years. Praise the Lord! Jean felt relieved and hopeful for the world to come.

She ran next door to show the photos to John. When she came back to show the photos to Mari she was jumping up and down in the hall holding on to her warmly clad bicep saying, "I was wrong about the neighbor! I was wrong! Yippee! I was wrong!" The two women laughed. "Would you be willing to take a ride into town?" Jean was asking Mari for company to drive into town to get the key back from the coroner's office.

A week had gone by since her neighbor had put his trash out for the last time. Jean remarked to Mari that trash day would be sorrowful for a long time to come. She was cleaning the hooves on her old white mare yet again, and was grateful she could still bend over to do so. She was happy to help.

Jean's daughter and son-in-law were making their plans to drive out West for the holiday season, and to arrive the day after Thanksgiving. They wanted to work remotely from Jean's house instead of their own house in North Carolina. Mari was asking Jean when she would have to move in next door with John to make room for the new comers. Jean was almost sure it wouldn't happen or so she hoped. Although, the governments—local, state and federal— had said not to do it, they were coming anyway. Jean finally arrived at a place in her head where she had accepted Mari's presence and was fine with the waiting game. They were enjoying nice adventures into town. Everyone was in a safe healthy daily routine, and then the game-changing passport finally arrived.

Jean hand delivered the mail to Mari. A couple of days went by, and then overnight everything changed. The governor of the State of Nevada issued a two week stay at home order unless a person had no choice. Abdul applied for Mari's visa in Turkey. Jean's daughter and son-in-law were on their way out west come hell or high water, and the dead man's sister was about to arrive to clean out his house and put it on the market. The three west coast states had issued travel advisories to alleviate unnecessary travel in and out, but Jean's daughter was bound and determined to get the fuck off the East Coast and breathe again without a mask. She was coming home to her mama.

Timing would all come together just as God would see fit. Jean didn't want for Mari to have to move in next door with John because all of her perfumed beauty products might send him to the hospital with respiratory failure. Jean had already developed a little cough because of the fragrances in the laundry soap. Jean's blood daughter would soon be arriving to meet her new black sister from another mother who would then soon be on her way to Istanbul to meet the man, who was hopefully-not-a-sex trafficker-or-organ-harvester. Jean was torn between her own selfish wishes of wanting to get back to her white privilege lifestyle, and the need to make sure that her BIPOC daughter was able to thrive in a despicable world.

She was cleaning up the messages on her phone when she found an unopened voice mail from the victim advocate person of Suffolk County asking her to return the call. That was Saturday in mid-November. Nothing could be done. Immediately Jean texted it to Mari who was at home next door at John's. Mari wondered why they had contacted Jean. "Let me think on it," Mari replied. She was feeling her own power. She knew Jean had signed off on the project. Jean felt slightly shocked for thirty seconds when Mari told that Abdul was now calling the shots, and that he would be deciding from here on out how to play the game of Cedric's Justice. This was the name Jean had given the project which she wanted to escape from as soon as God saw fit to let her service wane.

It was Sunday. John and Jean had just returned from church and a food supply run to Costco. Jean greeted Mari. They were always on good terms because they had a very succinct unspoken common goal: Get Mari Happily Married and Jean Off The Hook. Jean asked, "Mari, do you know why the DA's office called me instead of you?" Mari answered, "Because they lost my phone number?" "No! They called me because I was the squeaky wheel, the thorn in their side, the person that shoved sharp objects in dark places and refused to stay silent. I am the one who kept calling them, not you!" She continued, "All I've done is keep it open for you for when you are ready to move forward. I have done my job. You are the only one who can jump on it." Jean asked,"Have you told Abdul they called?" Mari replied, "Yes, I did. At first he said No, don't call her back!" He wants us to wait until we can be together." Jean wasn't surprised. Mari added, "And then he said, "okay, and that I could call to find out what she wants."

Mari asked Jean to call the woman on Monday morning at a reasonable hour so as to not look overly eager. The woman from the Victim Witness Advocate Office immediately recognized her name. "Hi, I'm Jean Lloyd. You called my phone on Friday and left a message." "Yes, yes, thank you for returning my call."

Jean fully intended to keep her mouth shut and only saying that she had Mari's permission to give out her contact information. But as usual Jean couldn't just play it cool and easy, and they got into it over the phone. Jean asked the woman if she were black or white and the woman refused to answer the question. Jean said, "That's cool. No problem. I'm white and I'm tired of the United States of America being a racist nation." The woman was not going to react. She was a highly trained professional.

Jean went on to say that Mari was scared to death of the system and that's why she had never made the contact. Jean continued, "The United States of America is world famous for being a racist nation, certainly you know that. It's all over the news every day!" The woman refused to let on that she had any knowledge. Jean continued, "Mari and I sat at the dining room table for well over three months taking down her statement. It wasn't easy to pull it out of her." She refused to admit it was in her computer for fear they would take remote control and destroy the tale of woe and incrimination. Then the woman asked, "Does Mari know that she has an attorney?"— "She does?" responded Jean.— "Yes, he was appointed by the court." "Oh? No. She has no idea she has an attorney." Then as if to play devil's advocate Jean asked, "Can she choose another attorney or does she have to keep that one?" Jean was playing dumb on purpose to see if she could get the woman to play her hand.

Jean detected the slightest degree of hesitation in the woman's voice, as if the woman was tempted to lie, but then she caught herself mid breath, "Mari is welcome to hire or fire anyone she chooses." Jean wondered if the advocate walked a fine line between the City of Boston, which must have been her employer, and her job as advocate for victims. The fact that the city of Boston had already obtained an attorney for Mari was an admission of guilt as far as Jean was concerned. If it had been left to Mari alone by herself nothing would have ever transpired. It was that vicious white dog Jean Stinson who was ripping the meat off the bone. Jean would tell poor old Mari, "You got black for soul. You got Mexican for strength. And now we're going to give you just a little slice of white in your pie to put that white fighting style into your backbone.

Don't put too much white in your pie though because I heard on NPR from some black guy that white people are insane. Jean was repeating herself, but every time Mari heard her say it she laughed. Laughing was good for everyone during these terrible times.

Jean decided not to text Mari the results of the phone conversation at work. Instead she waited for her to get home so they could talk without cell phones for fear of being bugged and recorded. She was convinced that the- powers-that-be were eavesdropping. They were both afraid of white power. But now it was up to Abdul and Mari. Jean asked her, "Have you told Abdul that our project, Cedric's Justice, is worth six million dollars?" Mari responded, "No. We are going to wait for us both to be together before I tell him the whole story." "Okay. Sounds good." Jean was excited to see what the young couple would do.

The advocate said she would give the contact information to Mari's attorney. Jean took down the name of the attorney, the address and the phone number and told the woman she would pass it on to Mari when she got home from work. Now the questions were: would Mari call her court-appointed attorney before she left for Istanbul? Would her court-appointed attorney reach out to her? Would Mari tell Jean if the attorney made contact? Time was of the essence. It was looking like Mari would be leaving in the next few days. She said she thought she was in her last week of work at the hospital. Abdul was telling her to quit her job again. There were so many variables and so many balls in the air.

When Mari arrived home that afternoon, Jean met her in the kitchen next door with the news about speaking to the Victim's Advocate. Jean was never good at mincing her words. If she had something to say she just let loose with it. It was not in her nature to pussy foot around much to Mari's chagrin. She had to dump it all out there, fast and furious. Mari would've much preferred her benefactor to use a little tact occasionally, but there was no time left. Jean had tried desperately to get off of caffeine for years. She had done real good for the last couple of days but this day she decided to have a cup of coffee about an hour before Mari got home from work so she had a major head of steam going on. Mari was a victim in more ways than one.

Jean started out fast and hard. If Jean had been in charge of Cedric's Justice she would have had an entirely different demeanor. Abdul was in charge now. He was the man that Mari had never met in person, but now he was in charge.

At this place in time the raging pandemic was claiming ten percent of all hospital workers. That was the report on the local news. If Mari could get herself over to Istanbul with Abdul, then that would make way for Jean's real daughter and entourage to move in for six weeks which she was not looking forward to because the stay-at-home orders meant for people to stay at home. That poor old white woman felt like she was living on the edge.

Jean ungracefully dumped her load of information on the ears of the poor young widow. "Mari, did you know that you have an attorney?"—"No."—"That's what I thought. I asked the woman if you had to keep the court-appointed attorney or if you could get a new one." Those parts of the conversation all went pretty civilized. But after a couple minutes the caffeine was pumping through her veins and Jean started in like she shouldn't have, but often did. She was using four letter expletives to get the point across as fast as possible. She wanted Mari to have the information to maximize her success. The young black widow, who had been thrust into the only living quarters she could find that were safe and free of charge, endured Jean's rant.

"As I see it, Mari, you have two choices. You can go with this court- appointed attorney and see what they offer, or you can change the world by making an example of Boston PD. If you go for the big bucks you will put the burden on the taxpayer. But if you stand on the principle of social justice reform and are willing to settle for less money in exchange for Cedric becoming an iconic name associated with social justice, perhaps Cedric's life really could matter." But Jean's logic was really of no interest to Mari. She played along just to keep Jean in service mode. She only had a few days left to go, and then she would be free and off to her new life.

Jean used the four letter words because she really wanted to get back outside with John to rake the autumn leaves. "You can go for the big bucks, Mari, and there is nothing wrong with that. But you have a chance to stand up for Jesus, and decency, which is exactly what those white motherfuckers need. They need to get their motherfucking asses kicked." Mari had become somewhat hardened to Jean's foul mouth, but she was still grossly offended by her crass outbursts, and Jean just wanted to get her point across. Mari was terribly offended that her husband's life had been reduced down to a dollar amount. Mari retreated into her new bedroom over at John's house around noon and didn't come out again until the following day. She was hiding and not eating. She was down to skin and bones.

She wanted to be thin to meet her new man and the time was growing near. Jean thought about how difficult it must be for a young woman of color to muddle through the choices ahead with a brand new husband. Mari and Abdul had practically the exact same skin color. That must have been a big part of the attraction. She hoped she hadn't pushed Mari too hard. Mari had little to no fight left inside of her. She just wanted a baby. Jean tried to explain to the advocate lady how much fear the widow victim had. It was not clear if it was falling on deaf white ears or ears of compassion. She had no idea if Boston was aware that it had a problem and wanted to become a better community, or if they just wanted to hold on to their white power forever. Jean was grateful to Abdul for stepping up and taking over the burden of Mari and her dead husband, but what if Abdul was only looking for American citizenship instead? What a good question!

# CHAPTER TWELVE

# Fear Of The Deal

Jean and John were out raking the leaves on the driveway. She was telling him about the conversation she'd just had in the kitchen. "I told her she has two choices and that she needs to make a commitment to whatever choice she makes." John was listening intently because he depended on his ex-wife for information. He was the stay-at-home-guy, and she was Mrs. Busy Body, a name he called her regularly. "She can either make a commitment to go for the highest dollar amount or she can…" She hesitated with her words, and John jumped in to finish the sentence, "go for principle." "Exactly!" Jean responded with gusto. "I told Mari she has, in the palm of her hand, the entire fate of mankind." John laughed and then he scoffed, "Go for the money. She should definitely go for the money!" Jean muttered under her breath, "Typical white male."

When she told the DA woman that she had taken Mari's statement over a three-month period the woman had immediately chimed back with Mari's brand new attorney contact information. It was like she was protecting herself from blowing the case for the city of Boston. But it could've been she was protecting Mari because, after all, that was her job. All three women knew Mari's statement was down in writing, but maybe it didn't matter. Jean decided she had better let Mari have a break from her obsessive white energy. She thought maybe she had pushed too hard.

After getting up the next morning she was resigned to letting Mari relax for a couple of days so she and Abdul would have a chance to chat. Things were tough so she wanted to give the young couple space. She'd done her job, and Abdul was taking over. Jean was in her kitchen washing dishes. Mari arrived clad in her multi-layered sleeping attire and fuzzy house slippers. The thermostat was still set at fifty degrees with snow on the ground outside. Mari jumped right in with both feet. She began with the advocate woman and the attorney. Jean was shocked but acted nonchalant. Mari was not wounded. She was playing in the game strong and resilient. This was a pleasant surprise.

They sorted out and shared the details that both would bring to the table. If Mari went with the court-appointed attorney, there was a reasonably good chance that person was sleeping in the same bed as the city of Boston. If she were to get a new attorney, who would that be? The conversation kept going back and forth. Mari clearly remembered her experience with that exact same court-appointed attorney just after Cedric's death. It was that very same attorney, who was still on the job. "I signed a document," said Mari— "What did it say? Do you still have it?" — "Yes. I think it's with Cedric's papers."

It was a simple document that said the attorney would try to negotiate a deal with Boston PD. It said that Mari agreed to let him hunt around and find out whatever he could in order to cut a deal. It also said he was not agreeing to represent her. The problem was that Mari did not like the way he had treated her at the time. Cedric was murdered in his own home on September 29th, 2019 before the Black Lives Matter summer of 2020 had changed the world forever.

The attorney, according to Mari, tried to convince her that her husband didn't matter. His words indicated that Boston did things differently. This was an important element in Mari's fear factor, and one of the biggest reasons she shut down, and refused to trust the government.

Jean believed that Jesus put helpful humanitarian thoughts into her brain that would help Mari. When she thought of the next step she always gave credit to Jesus even though she was a bit sour on Christianity. She had a hard time figuring out how the KKK was able to convince white Christians that the commandment Thou Shalt Not Kill did not apply to white folks. She Googled Civil Rights Boston. She had googled the same thing back in 2019 and gotten nowhere. Now there was a plethora of positive hits to choose from. She pushed the two buttons on her iPhone to take a screen shot, and texted it over to Mari. Neither woman wanted to make contact with the DA victim advocate woman, or the court-appointed attorney, but Jean desperately wanted to know the name of her son's murderer. I want to know if my son's murderer was Irish, that's all. She considered that to be her only paycheck. Jean was Irish and she wanted to know about her people. Did my people kill one of my children? I just want to know. Jean couldn't let it go. She was a lot like her Irish mother.

The old white mama kept coming back to Mari over the course of the next hour with various ideas while Mari was getting ready for work. Jean remembered that Breonna Taylor's family got the big money to start with and then, practically the next day, they turned the case over to the FBI as a civil rights investigation into the Louisville, Kentucky Police Department. Jean, still feeling guilty about selling her white brothers and sisters down the river, responded silently to herself. It is just too damn bad that they have to learn the hard way how to be decent human beings on planet earth. So fuck'em. There is only one way to turn this bus around and that's to go straight for the juggler. Jean and her white horse were back in the saddle again.

She told the young-black-widowed-pseudo-daughter, "Okay, I'm back on until you and Abdul can get together in person." That was her way of letting Mari know she wasn't letting go just yet. In Jean's mind it was a form of asking permission to continue. If Mari acted standoffish, Jean would have backed off, but Mari didn't, so Jean was back sitting high in the saddle riding western, and holding on to the horn for dear life.

All she really wanted was for the Earth to be able to heal itself. As long as human beings were still spending time on silly things like skin color the restoration of planet earth was not a happening deal. It would require every man, woman and child to fall in love with the planet, simple as that. The argument that the white people made the mess and the white people should clean it up doesn't hold a lot of water because many hands make small work. Everyone should be offered a job in restoration work if they are able. The problem was capitalism. Capitalism created soft, squishy, unhealthy lifestyles for people who subscribe to modern conveniences like trash compactors and garbage disposals. People are no longer strong and healthy enough to go out and work a hard day of re-planting forests. Asking the people of color to fix the white man's mess was going to be a hard sell. If the people of color were to become stronger and more healthy than the white folks then maybe the white people would start to value strong and healthy bodies instead of always trying to get people of color to do their work for them. Getting someone else to work only makes people weak. Jean felt everyone should want to work shoulder to shoulder to get strong and healthy together. The days of a Monsanto-run planet had officially been put on notice. Mankind would be returning to a pesticide free agrarian species or go extinct. Black folks and people of color are in the perfect position to save the human species from extinction. It was clear to Jean there was a good chance that the

BIPOCs would save the world from the evil white capitalistic imperialism.

This was no problem for old white Jean. She'd had a good life. She'd done her part to make the world a better place. She had welcomed people of color with open arms. When she had to sign up for her new Medicare doctor she asked for a young woman of color. She had the luxury of good health because she worked physically hard her entire life and not just sat behind a screen pushing buttons. She had much better muscle tone at sixty-five than Mari at age thirty-six. Jean's problem with white people was that a certain small segment of the white population held the rest of the world hostage. The world was in a stranglehold based on keeping people down so that others could run amuck on top. Jean saw that the haves and the have-nots were now being compromised into creating a better world together because of the pandemic. Jean loved Covid and saw it as a gift from God. It was nature's way of thinning the herd! Covid was Earth's best friend.

The topic of reparations, the one that the white folks prefer not to discuss, could be called a sore subject. Maybe the white people are afraid they are going to have to pay. The obsession with money is killing both the human race and the planet at the same time. Every step we take has a carbon footprint attached. After waking up to her divine purpose Jean chose to match her lifestyle to the goal of healing the planet. If the people of color could forgive the white people and if the white people apologized, then maybe we could all work together and get the job done. It seemed really simple to Jean. The problem was that the work force was no longer strong enough to work. The only healthy race Jean knew were Mexicans, and it appeared they were catching up to the modern world fast.

White people have been smart, there's no doubt about that. In their damming up of the rivers and streams to create electricity in their lust for power and harnessed energy (so people can pay electric bills) the areas on Earth called WETLANDS have been destroyed. These areas, which are also called riparian zones, are the biological filtering system for water on planet Earth. Plants are able to clean water by feeding on our filth and trapping sediment. Plants clean up our messes. This is how Earth keeps water clean. Because white people have been so aggressive with development and making money, life on Earth is on the verge of extinction. Jean prayed, to the Great Spirit, that the entire planet be healed before she died!

There must be many theories about why white people are the way they are as John who had a degree in biology told Jean."The first humans began migrating north out of the equatorial Rift Valley in Africa. There was a lot less edible vegetation up north because of the colder winters. They became meat- eaters. No more bananas. The further north they went the more meat they ate and eventually over thousands of years the high protein consumption grew a more aggressive brain." When Jean heard this a light bulb turned on. She was always trying to figure out how black people had allowed the atrocities to occur.

Many learned during Covid that not caring for people puts everyone at risk. John and Jean went into town to get some groceries before the Thanksgiving rush. They were talking while he was driving his cherished truck, which he considered a classic and not beat up. He stated that the South was against giving blacks health care coverage. Jean found that to be a curious thing and wondered if the whites in the South were afraid of the blacks because they knew they had it coming. Or were the whites afraid of the blacks for some other reason?

# CHAPTER THIRTEEN

# Bloody Thanksgiving

Across the street the dead neighbor's sister and sister-in-law were cleaning up the mess all week long. They used the brother's truck to make numerous trips to the dump. They visited the attorney and priest, arranged for furnishings to disappear, and sold the house in six days. Jean and the dead man's sister texted back and forth, but had never met face to face because of Covid. The time had come for a neighborly visit because the week was coming to an end. Mari, on early schedule at work that day, got home from the hospital in the afternoon. It was another beautiful autumn day on the backside of the Sierras.

Jean met Mari in the hall. "Hey Mari, would you want to do me a favor?"—"Yes! Anything!" She jumped at the chance to help. "You know the people across the street where the neighbor died?"—"Yeah"—"I need to go over there and say hello because they leave tomorrow and I just wanted to introduce myself. Will you come with me?" Mari loved the idea, "Sure! No problem!" —"Awesome, thank you."

While walking across the street Old Mrs. Jean Busy Body spoke to Mari of the wife of the sister being black, and that perhaps she felt a little intimidated by the grossly white neighborhood. She admitted to Mari she wanted to use her for her blackness. "Not a problem," said Mari. "Use away." This was a significant change from the Mari and Jean back in March at the start of the pandemic. In the beginning Jean had tried, gently, to get Mari to pitch in with chores, but always had the feeling that the young woman was resisting the idea out of fear or conditioning. Becoming a slave to a white person in a situation that was less than optimal simply did not appeal to Mari. Jean chose to back off and gave up asking for help until the car loan factored in to the equation. For Jean to ask Mari for a favor was a big deal! She only did it so the wife of color would feel more comfortable. There was no work required, only loving kindness. Mari believed in kindness, but it was clear she didn't believe in becoming a slave to a white person even if the white person was giving her a free place to live and a privately re-financed car loan. Mari, by walking across the street, was paying back a favor just by being kind. Jean hoped Mari didn't think that she was being used to help make her look like a good little white person to the dead neighbor's kin folk.

Both women were genuinely happy to meet Mari and Jean. They wanted a photo of the mother-daughter combo to take back home to the South. Jean and Mari squealed in delight, "Oh goodie! We get to take a photo together!" They immediately threw their arms around each other and smiled for the birdie. It was cute! They had, for years, joked about looking like mother and daughter. Now they had the proof!

Mari regularly used a great expression that Jean loved. It was simply the word "STOP." She would draw the word out California Valley Girl style. It always made her Mama Jeannie crack up. With the pandemic still in control of the world, and with Mari not being sure about her departure, and with the blood daughter showing up to take over the newly remodeled house that nobody was supposed to be living in so it could get sold, and along with a million other things…STOP! Jean thought in her own mind. Enough already! Her belief that God provides was her only saving grace. All she had to do now was to show up, and God would take care of the rest.

Both women decided to put the attorney and the DA's advocate on the back burner because neither of them knew what to do. Jean realized there was a strong possibility that Mari could still be living next door with John until after Joe Biden took office in January. Abdul's Turkish visa which everyone was waiting for, was not coming through. Jean suggested again that Mari get herself down to one suitcase and one backpack and be ready to leave on an airplane at a moment's notice.

Jean contemplated. Should Mari's eye witness statement be read by the FBI or an attorney? Maybe it should go straight to Dr. Phil or Oprah? She had been wondering these questions for months. With no idea in mind she knew that God would do as God saw fit. All Jean had to do now was to show up in the moment. Just like her blood daughter would do the day after Thanksgiving with her entourage.

But, just when she thought they'd all be settling into a nice comfortable lull, the energy shifted again. Abdul felt there was a strong chance that Mari would fly out on Thanksgiving Day. She had a holiday off from the hospital and planned on going down the hill with Jean and John to visit a neighbor who was celebrating his 70th birthday, which was a miracle. The guy had been a heroine addict for years and on Methadone for thirty. Jean enjoyed him because her own brother had been dead for a decade and a half due to a prescription drug overdose. She liked exposing both of her daughters to unusual diversity. She felt it was good to see how other people live and struggle. There by the grace of God go us… Obvious to some but not to all.

Mari informed her the night before about the possible impending departure. "No problem. No problem whatsoever." Jean responded with joy that Mari's plan was coming together. "Riiight. I knew it would be fine, but I just wanted to ask to make sure." "Yep. Piece of Cake. We can do it!" Yikes. Thought Jean. It's really here. It's really happening!

With the plan they had been working on for months in full swing Mari rose bright and early the next morning to finish moving all of her stuff over to John's house. This allowed Jean time to set up Mari's old bathroom and bedroom for the newcomers, who would need to quarantine for ten days before socialization could occur. Mari pruned herself down to one suitcase, one backpack and one toothbrush.

They met in the kitchen. Maybe it would be the last time. Mari had grown to rely on the old lady and trust was easy now. She appreciated that Mama Jean was never letting go. Abdul, on the other hand, had had so many sharp barbs throughout his life that he could not believe there was such a thing as a good person who wanted nothing in return other than a better world. He wanted to get Mari away from the white sugar mama so she could no longer be that influence on their lives.

Mari brought up the subject of the Jeep. Jean had paid it off and Mari was making payments back to Jean. The original car payment amount had been reduced by $300.00 a month. The new young couple were not sure if they should sell it, keep it, cut off the insurance and registration, or if Jean would volunteer to drive it back to the East Coast, where the couple would settle in NYC to raise a family. The plans were always changing. Abdul was no longer coming to the USA to retrieve his bride to be. Mari would be going to Istanbul instead. The possibilities were vast. Jean told Mari, "God will help you figure it out, and everything will be fine."

The African History Network on Facebook was a cyber-place where Jean liked spending time. She tried making all sorts of suggestions to the uninterested black people about how to be successful. She suggested they open up Chill University to teach white people how to take a chill pill and relax. Jean felt she had learned that skill from Mari. She also had some white friends that she would want to send off to the university to learn the art of Chill. She offered that this particular institution of higher learning could easily be set up in a reparation zone like Flint, Michigan, where her white Jewish- sympathizing, grandfather had been lynched by the KKK. She wanted to start a scholarship program at the new university if the black people would allow for white donations.

She told Mari about her suggestion to the African History Network and all she could say was, "Stop!" They roared in laugher. The inside joke became that whenever a scholarship candidate surfaced that person would be eligible for a full ride scholarship. Jean thought, Maybe Abdul needs a scholarship. She just let that thought roll on because she did not want to cultivate a negative vibe about Mari's new husband. She always said to her, "It's your turn, Mari, for good things to happen. No more bad."

As they were sharing the latest strategies of how to make everything streamlined, Mari hit a tearful snag, "What am I going to do with Cedric?" Jean was pretty good with that type of conversation, and stepped up to help the black widow by offering all sorts of suggestions like setting up an altar in her new room, or bringing him back to stay on the kitchen counter over at Jean's house. Jean didn't mind how Cedric's ashes would "hang out" while Mari was in Istanbul. She had an odd habit of believing in things lining up gracefully, and if that meant she would babysit Cedric's ashes for a couple of years, so be it. Cedric was not a problem.

Mari agreed, through a waterfall of tears, that Cedric would go to her new bedroom until she left for Turkey, and then come back to Jean for safe keeping. "No problem," Mama Jeannie replied. Mari was able to move on with her day of shifting completely to the other house before she had to be at work at one. But after Mari's cry, and they were still in the kitchen, Jean wanted to tell Mari about something that had happened during the middle of the night.

It was on her Ring Surveillance Camera app. She kept her cell phone next to the bed for charging. She had cameras installed up at the cabin because someone had started a car fire in front of the little house to be able to collect the insurance. The police told her Tahoe wasn't the same Tahoe as it had been back in the old days. She installed two cameras on the outside, and she put one camera on the inside to monitor the spirits because things would happen in the cabin that could not be explained. She got her first display of spirits on the night Mari moved Cedric's ashes.

"Check this out." Jean grabbed her phone and was pulling up the video showing two orbs floating upwards across the screen going from the floor to the rafters. "I captured my first spirit sightings on the surveillance cams." "No way!" responded Mari. "Seriously! Check this out." Jean suspected it was her family's spirits, but Mari saw Cedric in the 15 second video. Jean agreed, "Cedric! No doubt."

The blood daughter and son-in-law were arriving the next day. Jean had to get things ready for their arrival. She needed to put her house back in order by cleaning the floors which were already clean, hanging some fresh art here and there, and just generally enjoying her house all to herself before the next wave moved in to take over. She was preparing for setting up a campground out on the driveway and sleeping in her van. Jean wanted to become a snow camper anyway, so this was her window of opportunity to practice in a safe environment. Mari was mortified and felt like she was the problem, which was not the case. Jean liked opportunities to play, especially outside.

The problem was her own rule-breaking-white-privileged-children who couldn't do as they had been asked to do—to stay home during the holidays to get the pandemic under control. Jean was giving up her house to her own flesh and blood. There was always lemonade to be had when a lemon gets squeezed. The intrusion of Jean's real daughter forced Mari into one suitcase and one backpack and a move next door—out of Jean's house and into John's. Praise the Lord! Jean whispered to her inner thoughts when she woke up to just herself that first morning alone in nine months. Good things come to those who wait.

But the biggest conversation Mari and Jean were having was in reference to Mari's Jeep. Abdul seemed to think that he was in charge. Jean had to snicker at that one. Jean was holding the title for safe keeping. She offered all sorts of possible suggestions as to how to handle the situation. It appeared to Jean, that Big Man Abdul, who originally wanted to pay $9,000.00 for a first class airplane ticket, was now trying to get out of his future wife's car payment. They were considering taking the car to Carmax, getting cashed out, giving the money to Jean, and then making payments on the balance since the car value was under water. It was worth less than what they owed. Jean hated to see them lose the car because she felt it was a good deal for the new young couple starting out, but that was back when Abdul was showing up. It had flipped back and forth so many times, Jean gave up keeping track and just let go. What will be will be.

As Mari finished moving the last bits into the storage shed, Jean cleaned Mari's new bathroom in John's house and then got her the password for the new internet connection. She wrote it down on a good sized piece of bright orange junk mail, card stock. The doors to Mari's new storage shed outside were open and Jean sensed she was talking to Abdul on the cell phone. Mari appeared to be relaxed and doing well. She had done a beautiful job with her organizing and was feeling positive about her life ahead. All of her ducks were quacking in unison. Jean just wanted to slip the bright orange card into her hand so she could be self-sufficient in her new living quarters. Since Mari had already mentioned to Jean the lack of internet connection, the orange card was being dropped off to the open shed door with Mari inside. She handed it in, and Mari accepted with a great smile of appreciation and a quick thank you. Jean didn't want to interrupt so just waived 'you're welcome' and walked off. As she was walking away she could hear Mari pleading for Abdul's forgiveness for the interruption. Jean just kept walking.

The next morning Mari was over for breakfast because her food in the refrigerator hadn't been moved next door yet. They started chatting it up in the kitchen. They were laughing and having fun. Finally both had reached a point where they were able to laugh about each other's color. Jean had gone on for months about the evil white people. This had helped to gain Mari's trust. Mari was starting to let loose with some of the funny peculiarities of colored folk as well. Jean commented about the loudness of black people because they both had grown to love being in the kitchen wooping it up, loud style, with stories and laughter. Jean was trying to recant a funny commercial on television. "Have you seen that commercial on TV where the black guy is yelling at his next-door-white-dude-neighbor just across the fence, and the white guy tells his black neighbor he doesn't have to yell, and the black guy says something like 'I can't help it, I'm black'? Mari hadn't seen it but it launched them into more dialog about how laughter was a method of survival for black people, and that black women especially held the responsibility of keeping everyone's spirits lifted through laughter. Jean would sorely miss Mari's laughter when she was gone. Even the prejudiced, neighbor woman across the street had grown to love the sound of Mari's laughter, which occasionally broke the reliable peacefulness of their newly-integrated hoodscape.

She recalled back to the time when Cedric had been alive and they were all living together in the dog-pee carpeted house. Jean was very subdued back then. Mari and Cedric were as well. Both women had grown much closer to each other since the murder. Both of their survivals were intertwined. "You know what's going to happen, Mari? Abdul's not going to let us talk to each other anymore once he gets his hooks into you. Once he gains control."—"Oh, no. He's not like that." But Jean feared for her friend, her black daughter, and she hoped and prayed she was wrong and that Mari was right. Old Mama Jean went on to say, "He got pissy with you on the phone yesterday when I gave you that orange internet password card. You know what I'm saying, right?" Jean snaked her head around Mari over the kitchen sink and got right into her face eyeball to eyeball and then backed off quick as a church mouse. She just wanted to emphasize her concern.

They talked about how Abdul was not allowed to abuse Mari, and that she needed to have a return flight for safety's sake. Jean could see that Mari was torn between being an independent woman who could do really well on her own without a man, and becoming an extension of a man's penis. Jean was old. Mari was young. Jean had lived her life, and Mari was beginning over. Jean was trying to build her friend's self-confidence so that when she did get over to Turkey with Abdul she could be safe and well cared for and not fall victim to abuse. "He is not allowed to abuse you!" she reiterated.

Jean, on the other hand, was torn between wanting Mari to be gone forever so she could get back to her normal life, and fearing that it would all fall through and that Mari would come limping back to her pseudo-home only to start over again. Abdul was now becoming a threat to Jean. She heard herself say to Mari, "I'm your ace in the hole." As soon as those words left her mouth she knew she had just fucked herself in her own ass. She was still addicted to riding in on her fucking white horse. SHOOT IT! She screamed at herself in her own head. Jean visualized a dead white horse at the side of a country road with she, herself standing next to it watching the blood drain out.

# CHAPTER FOURTEEN

# The Print Out

"Mari, would it help if I printed you out a copy of your eye witness statement of Cedric's murder and sent it off to Abdul?" Mari's jaw dropped. She was used to not speaking her mind to white people, but Jean had become the exception. After Jean had declared her position as Mari's ace in the hole, she expected Mari to say, "No, you've done enough already, and I'm grateful. Thank you." But she didn't. She agreed with Jean that she truly was her ace in the hole, but she clearly didn't like the story of her deceased husband being sent to Abdul!

John and Jean had agreed about a month before Thanksgiving that once Mari was gone there would be no coming back. Jean suggested sending the eye witness statement to Abdul as a way to keep the chock behind the wheel to prevent further backsliding. She was keeping Mari's departure moving forward. She gave mixed messages to Mari. "Oh I love you and want the very best for you. That's why I'm always trying to give you as much information as possible to help prepare you for the nasty world. And by the way, when will you be leaving?" There was a constant push and hold, push and hold, push and hold. This was Jean's technique to maintain her upper hand with the situation. She wasn't going to kick Mari out into the streets, but she was praying like hell that she would soon be married and pregnant in Istanbul. That would solve everything.

Getting the house ready for the next wave to come through was almost complete. She didn't want to share her house with Jesus Christ, much less with her own daughter who was nowhere near as fussy about the kitchen sink as Mari had proven to be. Jean was going to post stickies all over the house. 'No dog on fuzzy white rug'. 'Squeegee the shower after every use'. 'Wipe down the kitchen sink'. Basics for some, but not for all. Jean was obsessed with her kitchen sink just like her mother had been. There was something about a pristine kitchen sink that gave a sense of protection against water borne diseases. She had never asked Mari if she flushed female products down the toilet, but she would be having this dialog with both of her daughters soon.

To give up her house to the quarantined situation meant she would not be able to make her daily inspections of the newcomer habits. She would be confined to her master-bedroom-converted-art-studio with the master bath. The new invaders would have the rest of the house, including the kitchen sink and laundry room. Jean would park her van close to the art studio so she could sleep outside but still have access to the toilet. She found that if she could do something unusual and out of the norm it helped her to practice psychological distancing. She could escape.

Mari was acting hurt or wounded about something, but Jean was far less concerned than she used to be. Perhaps Jean had shot her mouth off too much, or Mari was put off with having to move out. Or it might have been that Jean had put Mari and Cedric's platform rocking chair in the living room, where the new invaders might sit in it. Something was a miss and Jean felt the negative vibe. She traced back through her comments and wondered if it was something she had said, or maybe Abdul was being weird with her on the phone. There was always this sort of feeling that Mari only laughed at Jean's jokes just to maintain a safe place to live and to get help with her car payment.

The young black widow again told of how her parents were living in a small one-bedroom apartment with four adults, two of whom were Mari's sister and brother. That was the only way the people in her family could survive financially. Jean, on the other hand, was concerned that when the new invaders arrived, all the players would have their own bathroom. She worked hard all day just to get everybody ready for the next onslaught. She wondered if her comments to Mari about her ex-Italian-cop-boyfriend had been offensive. Jean, John and Mari were all over at Mari's new house. It was the first time they were all hanging out together at the kitchen table. Each participant was secretly aware of the magic in the moment. They had jumped the first hurdle.

That particular house—John's house— had seen its fair share of partying and richness of life events. It was not like the house next door. It was, now, a quiet house of easy chill and relaxation and not a house of tension. Jean said, "Mari, it's not a bad thing for everybody to discuss each other's cultural unique-ness." She was beating around the bush over the comment she had made earlier about her ex-Italian BF. "We need to be able to relax with each other and laugh at each other's genetic characteristics." She went on to say, "In Great Britain, there are four unique countries: Scotland, Ireland, England and Wales. They all hate each other and they're all white people." Mari raised her eyebrows, "Oh dear, that sounds about right when you're talking about white folks," Mari muttered this under her breath. Jean went on to tell about how the Scots hate the English, and the Irish hate the Scots and the English, and the Welsh must hate all three. Although she wasn't completely sure about the Welsh. It was normal to hate. That's how humans do it. Jean thought to herself, but went on to say, "If we can laugh at ourselves, pick apart all the bits and pieces, and fall in love with the miracle of God's creation then we DO stand a chance at evolution!" Diversity is Earth's only hope at survival. Diversity is the keeper of balance. Any scientist would agree. Jean's repetition was growing old, just like Jean.

She was, perhaps, overly obsessed with the fragile state of the world. She was sick to death of the Proud Boy mentality wreaking havoc for everyone. Excuse me Proud Boys, but you're pissing me off. The new established trio in the mellow party house next door was done for the day. They solidified their plans for Thanksgiving. Mari and Mama Jean would do the cooking, and John would be John, which was what he did best. Mari's Thanksgiving Day departure for Istanbul had been put on hold. Everyone retired to their respective rooms unlike Mari's family, who could not escape their close quarters. Jean was wondering about people of color having better manners than white people because of being forced into tight quarters with each other.

On Thanksgiving morning Jean watched the CBS morning news from bed. As usual Gayle King was the anchor. She had reported on the Black Lives Matter movement all summer long and into the fall. Holiday replacement newscasters were filling in for the regulars. The newbies were all people of color. No white faces were on the news that day for the white people to wake up to on Thanksgiving morning, 2020. Jean wondered if the Proud Boys were seething. It only made sense that they would be uncomfortable with their nation being usurped by people of color. All you saw on the TV for the last half of the year was black people. Jean wondered if that was just another whitewash or if black lives were really starting to matter!

John's friend up in the Pacific Northwest was a highly successful white male. He was also a Jew. John was telling Jean about an ongoing conversation the two white males had been having since the BLM riots in Portland had taken over the city for a hundred days that summer. John's friend stated he was unaware that black people had been treated so poorly. When Jean heard this she went slightly ballistic. She raised her voice to John and told him that his friend must be awfully stupid, which she knew wasn't true. She was just letting off steam.

Back in the kitchen the non-blood mother-daughter team were cooking up a storm. Now there was a new kitchen to divulge matters of the heart along with the daily nuts and bolts. Somehow having a new third party man on board made things even more stable. Jean was asking Mari one of her what-is- it-like-to-be-black questions. They got into a deep discussion over an episode that had occurred back in the early spring when Jean asked Mari to help wash the windows outside. Back then, the old white woman was afraid to ask for help because she did not want to appear to be a white slaver. She was afraid of Mari's blackness. But she also felt it was good for Mari to be required to pitch in, and to help out, since Jean was providing her with a free ride. The conversation continued because Jean continued it.

Jean told the story of how she had waited for the perfect window washing day with the perfect temperature and no wind. This was back in May. Mari had not yet started her new job at the hospital. It would only take an hour to do all the windows in the house if they could work as a team. All Mari had to do was point the hose at the window and Jean would do all the work. Slam dunk. No big deal. But there was this vibe going on with the two women, way back when, that Jean didn't understand, so she was asking Mari about it in the kitchen on Thanksgiving Day over prepping Brussel Sprouts for the oven.

"Remember that time you were helping me wash the windows?" Jean asked. "Yeah," and Mari got a big warm smile.

"Were you afraid that I was going to become a nasty white slaver and start bossing you around all the time?"—"Yeah, pretty much."—"That's what I thought." Mari went on to tell how she had a white friend once that was really nice and then all of a sudden she flipped out and got all nasty. Mari was trying to explain her position while Jean was remembering she had said to her once during the summer, "The only nigger and Mexican that lives in this house is me!" But she said it in such a way that no offense was taken or given. It was just Jean having fun, and Mari laughing at her inappropriate humor. Jean loved being as strong as an illegal Mexican laborer. She felt sorry for princess Mari who was weak physically, and whose diet was largely processed foods.

The new trio sat down to the table to eat their Thanksgiving meal together and Jean said, "Let's all say a prayer. Who will start? John I nominate you." John gave a lovely heartfelt prayer with no humor or cynicism. Mari was next. Her prayer was the standard Dear Father, We Thank You for bla bla bla, and Jean came in last with her standard 'For What We Are About To Receive May The Lord etc.etc.' The three human beings were chill with each other. Jean could relax over in that house, whereas in the other house all she could see was her investment dollars being destroyed before she had a chance to get the place sold. Jean's real prayer was that her blood daughter and entourage wouldn't mess up the new floors.

After dinner they made a quick dash down to their heroine addict neighbor friend's house to sing 'Happy Birthday,' for his seventieth. Nobody thought he would live that long. The plan was they would jump out of the car, go to the door to sing happy birthday, and then be back home in five minutes. Jean wanted Mari to have a chance to see white people in a pitiful state rather than spoiled and overly comfortable.

John, who was all freaked out about his asthma, said they would not be going into the toxic smoke filled house, but at the last minute changed his mind. The two women followed John into where the old heroine addict was planted in his overstuffed chair and dirty white bathrobe. The mother-daughter team stood together very close. They always stood shoulder to shoulder when things were weird. It was automatic. If one felt weird, shoulder up. If the other felt weird, the same. It was hard to say which one was more afraid of the world, the white or the black. Jean knew she had put her friend in a precarious situation and automatically assumed the position as her overly protective mother when she was asked by the wife of the heroine addict to introduce her daughter. "Yes! This is my daughter." She grabbed her and hugged her as if to say to the world DO NOT MESS WITH MY CHILD! PERIOD. It was very simple. Both women had a fierce Mama Bear for each other.

Jean didn't feel safe around white people until they could prove themselves trustworthy. She had too much invested in her science experiment of social integration to take the risk of somebody being an asshole to her child. She was very excited for Abdul to take over, but until he did she was still Mari's official watch dog. John was busy in the other room chatting with the wife, and Jean was busy chatting with the addict. There was only ten feet between the two. The addict was smoking a cigarette. Mari went to the other room, but the other conversation wasn't her cup of tea so she came back to Jean. The addict finished off the last puff of his cigarette then pulled out a dope pipe to light up. Jean heard Mari's muffled comment, "Oh my God, no!"

If it hadn't been for Cedric's last PTSD attack, the one that had been reportedly triggered by the smell of marijuana, Jean's alarm might not have gone off in her own head. "Okay, well then, we must be going." The addict was commenting about all of his friends who liked smoking pot. Mari moved into the other room. Jean gobbled down a small plate of the wife's raspberry cobbler just to be polite, put her fork in the sink, and off they went into the darkness of night. Jean said on the way back home, "Gee, I wish I could be a fly on the wall when you tell Abdul about this. That'll be a good one." They laughed, but none of it was funny. They got home and all three began their own kitchen clean-up. Mari wanted to take a piece of Jean's Thanksgiving vegan pizza with the spaghetti squash crust to work the next day. Mari's phone rang. Jean said, "I bet I know who that is!"

The day after Thanksgiving Mari had the late shift at the hospital. The unlikely trio, the three cohabitating humans, took a pleasant walk in the neighborhood before Mari had to leave for work. Jean texted her to say she had bits and pieces to drop off in her in-basket. This meant she had some new information to share.

She started by asking Mari if she had told Abdul about her visit the night before, and she had. "What did he say?" —"He said we certainly do have a lot going on around here, and 'I hope it's still like that when I get there'." Jean laughed and agreed, "Never a dull moment." The three continued walking. It was another gorgeous, crisp, autumn morning. Jean had a new idea she wanted to share with Mari. It involved sending Abdul a Christmas present. She felt strongly that Mari would not be leaving the United States before Christmas and that Mari could put together a gift to get into the mail ASAP so he would have something to open for Christmas Day.

Jean wanted to print out copies of her character witness statement, and Mari's eye witness testimony, wrap them up as a Christmas present, and send them off in the mail to Abdul. She could see the two documents making it in time for the holiday. Jean, Mari and John kept on walking down the street chatting. John was slightly put off that the conversation was not about him, but the women kept talking anyway. Mari was never rude. She was always polite. She did comment on Jean's idea, and said she wanted to wait until she and Abdul could be together face to face before she let the story out to her new husband in waiting. Jean agreed. She was just trying to move the process along like a good little honkey. Even though Mari did not want to send a copy of the two testimonies to Abdul, Jean went ahead and printed out both copies.

She had come to appreciate the reserve that people of color were able to exhibit. Because of white suppression always having the upper hand, she hoped when the BIPOC's gained equality they would not lose their humility only to become the same as the white folks. Chill University was still an option in Jean's mind. The black men could teach the white men how to chill. The black women could teach the white women how to chill. It could all be so much fun and everyone could learn to love each other. Why was that so difficult? The Earth was at stake. Without the Earth we don't have a planet to stand on! Jean ran this dialog in her head a thousand times a day. Her blood daughter would be showing up shortly. Jean was terribly embarrassed that her own flesh and blood was not able to follow the rules. It was so white. Jean had to move her food out, camp out, and live like a vagabond while her daughter took over, but Jean was very happy with the situation. A mother's love is always powerful. I wonder how much longer Mother Earth's love can sustain us?

Jean's blood daughter and son-in-law had never read the two testimonies. They had only heard the stories over the phone. The printer was in the art studio with WiFi capacity that could be signaled from Jean's bedroom. She had to load up the paper carrier before she could run the printer. She wanted the copies to be available for her real daughter to read during her stay, but she didn't want the story to leak out without Mari's permission. The two women still had no idea how the story would be told to the world. Would it go to an attorney, the FBI, the police, the DA? Jean had no idea what Mari and her new husband would decide. She only knew she wasn't going to let go of Cedric's Justice.

Back when Mari was getting rid of all of her stuff, Jean asked if the TV was going to the Salvation Army. It was. She asked if she could buy it from her, but Mari loved that she was able give it to her white mama instead. There was a video collection that Mari and Cedric had put together back when they were still in the military. There was a cute little white dresser unit on which the TV was perched, and on top of that a DVD player. They had purchased the little dresser unit as a kit in a small box. Cedric put it together with his tools from his bright yellow tool bag that always rode on the floor behind the driver's seat of their Jeep. Jean wanted to be able to put the two written testimonies in good envelopes and place them on the little white dresser for her real daughter and son-in-law. She went out to the garage and found two nice file folders in a box, one green and one red. Jean's logic chose to put Mari's statement in the red one because it was the same color as Cedric's blood.

# CHAPTER FIFTEEN

# The Statement

Mari Smith: Witness of alleged police misconduct and wife of Cedric Smith, who was killed by Boston PD, 9/29/19, Boston, MA

This statement was dictated to a close friend of Cedric and Mari Smith: Jean S. Lloyd

My husband began to have a PTSD attack, which began at approximately 6:50 pm on 9/28/19. It began when marijuana smoke blew into his face from another person present who was smoking pot. When I looked at him I could see my husband's face change and I knew he was going into an attack. I approached him and helped him walk away to the side entrance. We walked up the stairs and entered our apartment through the electronic-push-button-code locked door. He did not know where he was. It was like he was lost. He said, "How are you here with me?" It seemed like he was afraid I was with him on the battlefield in the war.

I was going to take him into our bedroom because I wanted him to feel safe, and I could see he was coming out of it. Two paying B&B customers who were staying with us at the time said hello to us. He was still coming out of it, and he said hello to them. My next thought was to get my phone so I could call an ambulance if it became necessary to call one. I never got my phone because I had left it downstairs on the table. Because I could not find my phone I came back out of the bedroom. When I got back he was missing. I ran up to see if he went upstairs, then I went downstairs. I followed the path outdoors that is the usual path because that was what I had been taught to do for him. I found him on the path.

He was hollering out saying, "Help me, help me, God! Someone call for help!" He saw me and knew who I was. He started to calm down but could barely walk. We were going slow. He was getting more confused again. He was calling out cadences from the military like calling out orders to soldiers and speaking Arabic and marching in a straight line. When he saw my face he stopped, and was okay again. We made it to the corner. He tripped on some unevenness and fell backwards. He hit the back of his head. He was begging for help. He was bleeding. It was not quite dark by this time.

I could see a person in the window across the street who had a phone and was indicating to me that a call was being made. To me, he looked very angry. Shortly thereafter the police turned up.

At that point, Cedric was standing up and we were starting to walk away. I could hear one officer saying "Do you see any domestic abuse going on here?" I heard the other officer reply after taking a look around, "No." The taller cop looked our way, then came over to us and said, "What the fuck is going on here?" The cop put his hand on Cedric's shoulder roughly, and demanded to know what was going on. I was trying to explain to the officer that my husband was having a PTSD attack. The cop said "Where do you live?" I responded to the officer, "That's our blue house over there."

Cedric and I were becoming more and more separated by the cops. The officer was trying to get Cedric to go over to our house, where we lived. Cedric kept telling the officer we were okay, and that we did not need any help.

I was telling the police all about the PTSD attacks and what my husband had done during the war. I was trying to tell them everything I knew about his background, his branch and his Military Occupational Speciality. The shorter cop was interested in hearing, but the taller cop was so aggressive toward Cedric that I saw my husband pushed into a deeper state of PTSD. I heard my husband tell the cop, "Look at what I have done for you, and look at how you treat me." By now we had entered our apartment through the locked door that I had unlocked. We were walking up the stairs in the stairwell. I could hear our friend tell the cop not to be putting his hands so roughly on Cedric.

We entered the kitchen on the second floor. The tall cop asked if we had any weapons in our room. Cedric and I both answered yes. I gave the officer the details of how our weapons were licensed and locked safely away. Cedric must have felt like we were in danger because he yelled for me to call the police. I kept asking the officers to please help us. My next request of the officer was to put my husband in handcuffs. I wanted him to be put in handcuffs so he couldn't hurt himself. I thought that if he could get put in handcuffs then it would enable all of us to help Cedric calm down. The shorter cop just gave up and walked away. I was begging the taller cop to put him in handcuffs so my husband could just calm down.

Then Cedric started backing up in fear toward our bedroom. As he did this, the cop was saying to me that my husband would not be able to open the electronically locked door because he was not in his right mind. Cedric began to push the buttons, just punching the buttons randomly, almost child like. I was so startled to see the door open. Cedric entered the room. I kept asking the cop if he could just handcuff my husband to keep him safe. Once Cedric was able to enter the room, I tried to go in with him and stop the door from closing, but I was unable to do so. I kept on asking the cop if he would help us and handcuff my husband to keep him safe. I could hear Cedric moving items around in our bedroom, and I became afraid for my husband's life. When I could see that the cop was not going to help my husband, but instead allowing him to access our firearms, I ran downstairs to get help. That is when I saw all the cop cars and the swat team. They were all yelling at me to get on the fucking floor. Then I heard a gunshot go off. I heard over the PA system the words "warning shot fired".

I could see Cedric in the window raising his hands yelling, "Help me." I moved to the side of the house to get away from the cops pointing guns. The cop who was in the back, by the gate, told me I had to leave and said he didn't care that I was Cedric's wife.

I walked around the other houses to get back to the apartment and I came across more police, so I tried to approach one of them to explain that I was his wife. I was trying to tell them that I was his wife, and to explain what was happening with Cedric regarding his PTSD and military service. The cops were going for their guns and telling me to raise my hands and to back the fuck up.

I raised my hands and started backing up slowly. When I got far enough away I came in contact with the man who said he was the negotiator. He then began asking me all sorts of questions in regards to Cedric.

He said that his job was to take me to where Cedric was so that I could calm him down, and told me not to worry. He said, "Don't worry, your husband is not going to get killed." And added that the last thing they would want to do was to kill a war veteran. He went on and on about how Cedric would get help.

About ten minutes after he talked to me I heard the second gun-shot. I was starting to have a panic attack. I saw that they had a homicide van parked at the scene. I thought that meant that they had killed my husband. I kept trying to get to him but they would not let me go to him. They were saying I should wait for the ambulance.

I kept asking about the homicide van and they wouldn't let me go to my husband. There were all these officers around me, so I couldn't get to my husband. They made me go to the ambulance and get my vital signs checked. After the ambulance, two negotiators told me they had strict orders to take me to the person in charge, who was just a short walk away. I asked if I could go to my husband first. They said no, I had to go to who was in charge instead, and that he would tell me everything.

Based on how they were talking to me I followed their orders and walked with them. We ended up not doing what they had said though, instead they took me to a city bus for civilians, who were people staying at the B&B where we lived, worked and managed the property, as well as some close neighbors affected by the incident. Before I went into the bus I kept asking the officer why I was getting on the bus and not talking to the person in charge. He told me not to leave the bus, and then he closed the doors. After he left I went to go find a spot to sit down. I could hear people talking and I heard a child ask its parent, "Why do they want to hurt that nice man?" The parent replied, "The police need to do their job." The child asked about the smoke coming from the home. The parent said, "That is a smoke bomb." The child said, "Doesn't that hurt people? I think it makes people sick."

My friend came to the bus and found me. She boarded the bus and asked, "How are you doing and why are you on this bus?" When she asked me the question I responded, "No, I am not okay, and I was told I was supposed to be taken to the person in charge, but instead they told me I was not allowed to leave this bus." She said, "But are you okay?" I replied, " I feel like I am starting to have a panic attack." She said, "They can't legally hold you on this bus."

My friend walked to the front to get help. She got off the bus and said to the cops, "Can somebody help me? My friend is having a panic attack." They asked, "Who is your friend?" I was able to hear her talking because I was in a seat close to the front of the bus. I was standing up by that time. I heard my friend tell the cops that I was the wife of the man that all this was about. She told the cops that it was illegal to keep me there, and that I was having a panic attack. I heard the cops say, "Oh no, she doesn't have to stay." So then I got off the bus.

The negotiator said for me to walk over and sit on a bench. I asked him if we were going to talk to the person in charge like he had said originally. Then he just left me sitting there. The second negotiator who came was a female. I asked her why I was being made to sit here. She said she didn't know. She was very apologetic about everything going on. Then two detectives in plain clothes came and said they would take me to the police station.

I told the lady negotiator that I was not comfortable going to the station because I didn't know those people at all. The lady said she would go along with me. Then two uniformed cops came with their police car and said they were there to take me to the station. My friend rode with us as well. The three of us rode in the back seat of the police vehicle with me in the middle seat position.

We arrived at the police station. We went upstairs. My friend was asking the other detective, "Is anyone going to tell us what is going on?" They wouldn't tell us what was going on with Cedric. They wanted us to give our statement, but they wouldn't tell us what was happening with Cedric. I kept asking for information about my husband. The detective was perturbed because he had been called in to the crime scene and woken up out of his sleep. He kept saying that he didn't know anything. They wanted me to give my statement, but they wouldn't tell me anything about my husband.

The detective asked, "Who wants to go first?" I volunteered. I asked him, "What are we doing here exactly?" He said for me to follow him into this room and that he had a lot of questions for me. Once we entered the room I asked for an update again. I asked to know if my husband was still alive. The detective responded about his sleep again, and that he knew nothing and that he has just been told to question us. I remember thinking how weird it was that he was supposed to question us but he knew nothing. I was really apprehensive.

He then said, "You should just start with giving me your statement about everything that's happened tonight." I told him, "I don't feel comfortable speaking until you tell me what has happened to my husband because I have cooperated all night long and you haven't told me one thing about how my husband is doing." I then looked at the lady negotiator in an attempt for help and she rubbed my back and told the man detective, "She has been very cooperative all night long and I think she is owed an explanation as to where her husband is right now, and what is going on. She has asked for information about her husband several times."

The detective looked at the female negotiator as if he were angry with this comment. He then said he'd be right back and was gone for 15-20 minutes. When he returned he responded with, "I was not able to get any information at this time." I responded, "I don't understand what's going on here. Something doesn't feel right. I will not be answering any questions until I find out what's going on with my husband." He left again.

While he was gone the female negotiator and I were talking. When he returned he said, "Would you be comfortable writing down your statement of what happened? I know you are more comfortable talking with the female negotiator." At that point I suspected him of lying about the closed circuit camera system being switched off since he knew that she and I were discussing things. At that point I did not say anything. I just looked at him and then I asked, "Did you get any further information about my husband?"

He then said to me, "Yes, I was able to find out. Your husband is no longer with us."

The rest of it all seems a bit blurry. I remember asking him what the time was of my husband's death? And then I asked, "Where is his body?" And then he said, "I don't have that information. That's all I have." He then said, "If you are not going to give your statement tonight, then let me give you my card and you can give it tomorrow." After that I asked, "What am I supposed to do now?" because I knew my house would be off limits.

# CHAPTER SIXTEEN

# Reparations and White Guilt

The printer was loaded with paper and ready to print out two recollections of events. Jean pushed print. Many of her friends had become Mari's friends also, and many white people knew the story but had not yet read Mari's testimony. It was now available if needed or wanted.

The young white privileged invaders arrived after dark in their newly purchased tricked out mini camper van that they had both worked hard to procure and develop. Jean was comfortably tucked away in her old camper van out on the driveway sleeping better than in the house. Mari was next door. John was always fine. For the next week or so the quarantine ahead would prove to be interesting. Jean wondered if Mari and her blood daughter would ever have a relationship. It was unlikely. Jean had mentioned to Mari when the trio was out walking the dog the day after the holiday, "What if we write a book, tell the whole story and it becomes a NY Times bestseller? Then we don't have to bother with the cops, the DA, the FBI or the attorneys. I don't want the money. You can have the money. First you can pay off your car and then you can buy a house." She went on to say, "That way the story can go straight to the American people." "I like that idea," said Mari.

Jean never knew what Mari was really thinking. There was an unspoken understanding that neither would ever be blood. Mari's expectation that Jean would turn on her at any time was a given. There was no way Mari could get past the fear that her white benefactor might snap. Jean on the other hand knew she was being used for her whiteness. Mari liked the way white people lived. Mari had become the princess in Jean's world and Jean was the slave. If only she could get back to her simple white privilege! Her problem of being desperate to be loved and always riding in on her white horse to save somebody's day was too much work. She just wanted her life back without the horse under her sorry white ass.

She was good at cutting backroom deals with God. "God, you get me out of this mess, and I promise I will mind my own business from here on out." Then she thought about reparations, the topic the white people were afraid to discuss. She was peeking over John's shoulder reading his email while giving him a hug from behind. She saw that the churches were having a meeting with multi-faith clergy. They were attempting reconciliation. How can reconciliation occur when white people refuse to apologize? Their standard statement is that their ancestors weren't slave owners, and that slavery had nothing to do with them. They are in their own little world, and America gives them the freedom to do whatever the fuck they want. Jean ranted on in her own head. Can humans be trusted with freedom?

It seemed pretty obvious to Jean that the American system needed a facelift. How would the privileged white people transition into caring about their country and the world instead of just themselves? What was it going to take to turn this bus around? In many ways the United States of America had failed at becoming a great nation because it was only great to a few and not to all. She thought it was ironic that the freedom that Cedric had fought to defend during the war was the same freedom that had put him to death, but that's the way it was in America in 2019. The Europeans didn't want to visit anymore because they were too afraid of getting shot. America was well known for being armed and dangerous. All Jean could do was shake her head in disgust.

What would the reparations look like? Jean pondered this question almost as much as how she was going to restore the entire planet before she died. Would the black people just get a big pile of money to spend however they wanted? Hadn't the white people done that same thing with the Indians who were now going extinct because of Covid? It seemed to Jean, who couldn't help but think like a white person, that education would be a good place to put reparation dollars to work. The problem was who would write the curriculum? The restoration of the inner cities might be next in line for reparations if everyone could agree on where people would go when their buildings are either being remodeled or torn down. The daunting task of reparations was insurmountable unless black folks and white folks could start talking to each other and work together. Learning how to talk to a person who was not the same color as you were the first challenge for the new America. Nothing good would come unless people could talk. It was a whole lot for old white Jean to think about.

Riding in on her white horse to save the black widow just happened out of the ether. It wasn't planned, it came from Jean's desperation to be loved. She had finally reached a point of being worn out. It was time for her to look after herself while she still could. She was getting old. She was glad she was able to help the young woman of color. She was grateful God had made her strong enough to help someone else, but now it was someone else's turn to help. Jean just wanted to get back to her white privilege. She no longer wanted to run her own social services agency. She felt she had done her part and now it was time to rest. Mari could never tell Jean how she truly felt about anything because she could never get past the fact that her benefactor was white, and white people could never be fully trusted.

# CHAPTER SEVENTEEN

# White Guilt

The systemic racism created by the abuse of power had finally reached a tipping point. The USA had to figure a way out of the white militia groups taking the law into their own hands. The United Nations envisions all people on Earth shall have a place to live, healthy food to eat, not live in fear of violence, children educated, a pollution-free world with economic opportunity, et cetera. The United States of America, the wealthiest nation in the world, is being asked to join the rest of world as a team player. The UN has come up with the seventeen Sustainable Development Goals for all of the nations of the world to follow for the purpose of evolution and planet restoration. How would America meet the requirements of the UNSDG's while at the same time protecting its precious freedom for the individual? Jean still wasn't letting go of the dream.

She had time, now that the Thanksgiving holiday was past and her two daughters were both safe and healthy. She wondered again if the two young women would ever meet. Everyone was still quarantining for a few days more. Did the two women even want to meet each other? There was no way to tell. Both women loved what Mom could do for them. There wasn't anything Jean's blood daughter couldn't do if she put her mind to it. She simply had no barriers to success. Mari, on the other hand, was afraid to relax and believe in good things to come. She couldn't. Jean must have told Mari a thousand times, "It's your turn Mari. It is YOUR turn!"

Mari wanted to believe in the new up-and-coming world for people of color. She wanted to think that her black child might have a chance at health and happiness, but that little seed in the back of her mind was always ready to pull the race card. It would never disappear.

Jean wondered about the power of an apology, if that could even stand a chance. White people would never apologize, and if they did the people of color wouldn't accept it anyway. No wonder Jean wanted to disappear back into her white privilege because she certainly couldn't do it all by herself. The same with the Earth— she couldn't repair the Earth all by herself.

Reparations would be a turning point for the human race. Repairing the damage done to each other and to the planet would become the future of all evolution. Some people believed in The Rapture and thought that the destruction was normal, and that only some would go to heaven in the end. Jean was okay with the extinction of mankind, but she hoped it would happen today instead of tomorrow. The sooner the better. She never stopped dreaming of the collective human consciousness, when all would work together as a species. She knew that the United Nations was working towards developing a human race that did not abuse power. That would be a very hard item to sell to the Proud Boys in their pirate trucks. They had the right to do whatever they wanted. They were Americans. They were living their dream.

Living on the edge is never easy. Jean got a slight taste of the edge by being so close to a woman of color. She said to Mari many times, "I would never have traded this experience for the world." There was always a feeling that Mari was grateful, but the resentment toward white skin would never go away. It would take ten generations to change the world. That's if we started today. Jesus brought the message of love, and do you think we have been able to get it right in the last two thousand years? Hardly. What possible chance was there for the world? Jean often reflected back on one of her mother's old expressions. "You can't legislate morality." Jean figured that meant you can't make it law for people to be kind and decent to one another. Her mother would also say, "People are no damn good!"

Jean was at the age now when it was tempting to give up hope for the world. Why should she care? She could go back to her white privileged ways and just enjoy the rest of the days she had left on earth. She would get involved with the various environmental groups and continue her bird watching hobby which she and John had been enjoying together over the past few years. She would continue to feed her dog better than most humans. She would continue to be obsessed with her new Apple Watch, the one that monitored her oxygen levels. She'd go up skiing at the cabin and hiking on the other side of the lake. She'd continue to have amazing health because she ate from her giant organic vegetable garden. She'd travel by train instead of plane after the pandemic passed. She'd hire an electrician to switch out some fixtures on one of her three houses. She'd sell one of her houses because she could. She'd pay extra for organic milk from cows that ate fresh grass. She'd continue to keep tabs on Mari, and be her friend because she loved her as if she were her own flesh and blood, even though she knew that blood was thicker than water.

There would always be good times and bad for every race, creed and color. Jean hoped that the white people would become kinder and more giving because she felt that the white militia groups were setting all white people up for a bad ride. Seven percent white people on Earth was clearly a minority. Jean couldn't understand how the white people could be so short-sighted and not realize the predicament they had created for themselves. She was tired of her own color being the aggressive color on Earth. She didn't necessarily want the white folks to become passive, but she thought it was a good idea for white people to park their greed at the back door and take a hint from all the black brothers and sisters. Learning how to chill would be a good lesson to learn. The blacks chill because they have no choice. The whites could learn how to chill because it might save life on Earth.

Nobody, or even the planet for that matter, was going to put up with flaming assholes forever. Consider this a warning. Either put the kind in mankind, or go extinct! Jean was cool either way. Things would get better for everyone, or extinction would prevail. The days of the industrial revolution were coming to a close. Disappearing are the people who hoard their wealth with no contribution to the greater good. Jean wondered if she'd be able to help someone else in the future, even if she didn't really want to. It's a privilege to be in a position to help. Maybe my white privilege doesn't matter anymore. Maybe it's gone forever. Better go out for a hike while I still can. Jean didn't have to think twice. At her age she realized walking was a privilege also.

The End. Or the beginning?